# Human Vaccines and Vaccination

# THE MEDICAL PERSPECTIVES SERIES

Advisors:

**D.R. Harper** *Department of Virology, Medical College of St Bartholomew's Hospital, London, U.K.*

**Andrew P. Read** *Department of Medical Genetics, University of Manchester, Manchester, U.K.*

**Robin Winter** *Institute of Child Health, London, U.K.*

Oncogenes and Tumor Suppressor Genes
Cytokines
The Human Genome
Autoimmunity
Genetic Engineering
Asthma
DNA Fingerprinting
Molecular Virology
HIV and AIDS
Human Vaccines and Vaccination

*Forthcoming titles:*

Antimicrobial Drug Action
Antiviral Therapy
Antibody Therapy

# Frontispiece

The official parchment certifying the global eradication of smallpox, 9 December 1979.

# Human Vaccines and Vaccination

*M. Mackett*

Department of Molecular Biology, Christie CRC Research Centre, Paterson Institute
for Cancer Research, Wilmslow Road, Manchester M20 9BX, U.K.

*and*

*J. D. Williamson*

Department of Medical Mirobiology, Wright Fleming Institute, St Mary's Hospital
Medical School, London W2 1PG, U.K.

*β*IOS
SCIENTIFIC
PUBLISHERS

© **BIOS Scientific Publishers Limited, 1995**

First published 1995

A CIP catalogue record for this book is available from the British Library.

ISBN 1 872748 77 5

**BIOS Scientific Publishers Ltd**
**9 Newtec Place, Magdalen Road, Oxford OX4 1RE, UK**
**Tel. +44 (0)1865 726286. Fax +44 (0)1865 246823**

*615.37*
*MAC*

DISTRIBUTORS

*Australia and New Zealand*
  DA Information Services
  648 Whitehorse Road, Mitcham
  Victoria 3132

*Singapore and South East Asia*
  Toppan Company (S) PTE Ltd
  38 Liu Fang Road, Jurong
  Singapore 2262

*India*
  Viva Books Private Limited
  4346/4C Ansari Road
  New Delhi 110 002

*USA and Canada*
  Books International Inc.
  PO Box 605, Herndon, VA 22070

*12424*

Typeset by Marksbury Typesetting Ltd, Midsomer Norton, Bath, UK.
Printed by Information Press Ltd, Oxford, UK.

# Contents

# Abbreviations

| | |
|---|---|
| AcMNPV | *Autoradiographica californica* multiple capsid nuclear polyhedrosis virus |
| ADA | adenosine deaminase |
| ADCC | antibody-dependent cell-mediated cytotoxicity |
| ADP | adenosine diphosphate |
| AIDS | acquired immunodeficiency syndrome |
| *amp* | ampicillin resistance gene |
| APC | antigen-presenting cells |
| ARR | avian reticuloendotheliosis retrovirus |
| BALT | bronchial-associated lymphoid tissue |
| BCG | bacille Calmette–Guérin |
| BL | Burkitt's lymphoma |
| BPV | bovine papillomavirus |
| BS-WC | B-subunit plus inactivated whole-cell |
| cAMP | cyclic adenosine monophosphate |
| CD | cluster designation |
| CHO | Chinese hamster ovary |
| CMI | cell-mediated immunity |
| CMV | cytomegalovirus |
| CNS | central nervous system |
| COS | see Glossary |
| C region | constant region |
| CSP | circumsporozoite protein |
| CTL | cytotoxic T lymphocytes |
| CVI | Children's Vaccine Initiative |
| DAF | decay accelerating factor |
| DDT | dichlorodiphenyl trichloroethane |
| DHFR | dihydrofolate reductase |
| DISC | disabled infection single cycle |
| DPT | diphtheria, pertussis and tetanus |
| EBNA | Epstein–Barr virus nuclear antigen |
| EBV | Epstein–Barr virus |
| ELISA | enzyme-linked immunosorbent assay |
| EPI | Expanded Program on Immunization |
| ETEC | enterotoxigenic *E. coli* |
| F | virion fusion protein |

| Fab | antigen-binding fragment of antibody |
| FB | Factor B |
| Fc | crystallizable antibody fragment |
| FD | Factor D |
| FDA | Food and Drug Administration |
| FP | fusion protein |
| $G_s$ | stimulatory G protein |
| GALT | gut-associated lymphoid tissue |
| HA | hemagglutinin |
| HBc | hepatitis B virus core protein |
| HBcAg | hepatitis B virus core antigen |
| HBs | hepatitis B virus surface protein |
| HBsAg | hepatitis B virus surface antigen |
| HBV | hepatitis B virus |
| HCMV | human cytomegalovirus |
| HCV | hepatitis C virus |
| HD | Hodgkin's disease |
| HgR | mercury resistance |
| Hib | *Haemophilus influenzae* serotype b |
| HIV | human immunodeficiency virus |
| HLA | human leukocyte antigen |
| HRF | homologous restriction factor |
| HSV | herpes simplex virus |
| HTLV-1 | human T-cell lymphotropic virus type 1 |
| IARC | International Agency for Research on Cancer |
| IBVD | infectious bursal disease virus |
| ICAM | intercellular adhesion molecule |
| Id | idiotype |
| IFN | interferon |
| Ig | immunoglobulin |
| IL | interleukin |
| IM | infectious mononucleosis |
| InC P | incompatibility group P |
| IPV | inactivated poliovirus vaccine |
| ISCOM | immunostimulatory complex |
| JEV | Japanese encephalitis virus |
| KLH | keyhole limpet hemocyanin |
| LAK | lymphokine-activated killer |
| LCL | lymphoblastoid cell lines |
| LFA | leukocyte functional antigen |
| LGL | large granular lymphocytes |
| LMP | latent membrane protein |
| LPS | lipopolysaccharide |

| | |
|---|---|
| LT | lymphotoxin |
| MA | membrane antigen |
| MAC | membrane attack complex |
| MALT | mucosa-associated lymphoid tissue |
| MDV | Marek's disease virus |
| MHC | major histocompatibility complex |
| MMR | measles, mumps and rubella |
| MOMP | major outer membrane protein |
| MTX | methotrexate |
| NA | neuraminidase |
| NADPH | nicotinamide adenine dinucleotide phosphate (reduced form) |
| NANA | *N*-acetyl-neuraminic acid |
| NDV | Newcastle disease virus |
| NIH | National Institutes of Health |
| NK | natural killer |
| NP | nucleoprotein |
| NPC | nasopharyngeal carcinoma |
| NYVAC | see Glossary |
| OPA | opacity-associated protein |
| OPV | oral poliovirus vaccine |
| ORF | open reading frame |
| P | properdin |
| PCR | polymerase chain reaction |
| PHC | primary hepatocellular carcinoma |
| PLDLG | poly(D,L-lactide-co-glycolide) |
| PRP | phosphodiester-linked polymer of ribose and ribitol |
| PrV | pseudorabies virus |
| PSH | pulmonary syndrome hantavirus |
| RMAS | see Glossary |
| RSV | respiratory syncytial virus |
| RT | reverse transcriptase |
| RT-PCR | reverse transcriptase polymerase chain reaction |
| SCID | severe combined immunodeficiency disease |
| SIV | simian immunodeficiency virus |
| SSPE | subacute sclerosing panencephalitis |
| STD | sexually transmitted disease |
| SV40 | simian virus 40 |
| TB | tuberculosis |
| $TCID_{50}$ | see Glossary |
| TCR | T-cell antigen receptor |
| Th | T helper cells |
| TIMP | tissue inhibitor metalloprotease |

| | |
|---|---|
| TK | thymidine kinase |
| TNF | tumor necrosis factor |
| UNICEF | United Nations Children's Fund |
| VCA | virus capsid antigen |
| VLP | virus-like particle |
| VP | virion protein |
| V region | variable region |
| VSG | variant surface glycoprotein |
| VZV | varicella-zoster virus |
| WHO | World Health Organization |

# Preface

The main aim of this book is to show how both immunological principles and molecular biological knowledge may be used together to generate new vaccines and to improve existing vaccines. The pathogenic mechanisms of infectious and parasitic organisms are described first, followed by the innate and adaptive immune responses they may provoke. Further sections then describe methods used for the identification of vaccine antigens, together with the application of new technologies for their expression and delivery, including recombinant live vectors. The molecular approach to the further development of existing attenuated killed vaccines and subunit vaccines is also described. In the past, vaccine development was seriously hindered if the pathogen could not be grown outside its natural host. However, recombinant DNA technology now allows the transfer of genetic information from fastidious organisms to more amenable hosts such as *Escherichia coli*, yeast or mammalian cells. This technology provides a novel means for the expression of protective antigens to be used as vaccines. Such is the power of molecular biology that even if it is possible to grow pathogens *in vitro*, recombinant DNA technology is being used to provide safer, more effective vaccines than those currently available. Finally, the book assesses progress towards vaccines for several key pathogens and addresses related issues such as the implementation of vaccination programs.

M. Mackett
J.D. Williamson

Chapter 1

# Communicable diseases: the problem and some solutions

## 1.1 An overview of communicable diseases

Infectious and parasitic diseases are the major causes of high mortality in developing countries, and are particularly responsible for perinatal and childhood deaths (*Table 1.1*). Global age-related statistics show more than 15% of all infants in developing countries will not reach the age of 15 years, whereas in developed countries less than 1.5% die in childhood.

**Table 1.1:** Life expectancy, number of survivors and chances of eventual death from infectious and parasitic diseases: examples from developing and developed countries

| Example | Age (years) | Eventual death from infectious and parasitic diseases (risk per 1000 cases) |
| --- | --- | --- |
| Developing country in WHO–African region | 0 | 195.9 |
| | 1 | 186.4 |
| | 15 | 149.0 |
| | 45 | 140.8 |
| | 65 | 120.1 |
| Developing country in WHO–Americas region | 0 | 68.6 |
| | 1 | 64.0 |
| | 15 | 60.9 |
| | 45 | 61.8 |
| | 65 | 61.7 |
| Developed country in WHO–European region | 0 | 4.2 |
| | 1 | 4.1 |
| | 15 | 3.9 |
| | 45 | 3.6 |
| | 65 | 3.2 |

Data reproduced from ref. [1] with permission from the World Health Organization.

Many communicable diseases (e.g. measles, tuberculosis, typhoid) are endemic, that is they constantly afflict certain human populations within particular geographical areas. Such diseases are often associated with economic deprivation in developing countries but they can flare up as epidemics in developed countries when the social infrastructure is disrupted by natural disasters or by war. Overcrowded living conditions hasten transmission of respiratory diseases through aerosols containing infectious droplets generated by coughing and sneezing. Water supplies can readily convey gastrointestinal diseases if they are contaminated with human sewage. The geographical distribution of other endemic diseases (e.g. malaria, yellow fever) is determined mainly by climatic factors. The temperature and rainfall typical of tropical or subtropical regions support abundant insect populations, especially blood-feeding insects, which transmit infectious and parasitic diseases between human hosts or between human and animal hosts.

Some communicable diseases (e.g. hepatitis B, acquired immunodeficiency syndrome (AIDS)) are pandemic, that is they occur on a worldwide basis. Clearly, unlike endemic diseases, their spread is not so readily restricted by environmental or climatic factors. Pandemic diseases tend to be transmitted either sexually or by the respiratory route. Their spread in recent years has been made easier by the huge increase in international trade, travel and tourism. Tetanus is a disease that may be contracted accidentally in almost any part of the world, although local customs or practices may significantly increase the risk of infection.

In 1990, deaths due to infectious and parasitic diseases accounted for 44% of total deaths in developing countries compared to 4.4% of total deaths in developed countries. The awful toll on human life exacted by particular endemic and pandemic communicable diseases is shown in *Table 1.2*.

More than 2 billion people (about 40% of the world's population) mainly in tropical and subtropical countries, live with a risk of exposure to malaria (*Figure 1.1*). This parasitic disease is transmitted by an insect vector, the *Anopheles* mosquito. About 100 million cases of clinical malaria are reported annually, leading to between 1 million and 2 million deaths each

**Table 1.2:** Global incidence of selected communicable diseases

| Disease | Annual episodes ($\times 10^6$) | Annual mortality ($\times 10^6$) | Case:fatality (%) |
|---|---|---|---|
| Malaria | 100 | 1.5 | 1.5 |
| Hepatitis B | 300 | a | a |
| Tuberculosis | 8 | 3 | 37.5 |
| Pertussis | 51 | 0.6 | 1.2 |
| Measles | 67 | 1.5 | 3 |
| Typhoid | 35 | 0.6 | 1.7 |
| Tetanus | 1.8 | 1.2 | 67.7 |
| HIV/AIDS | 0.32 | 0.26 | 80 |

[a]Significant number of deaths from chronic disease.

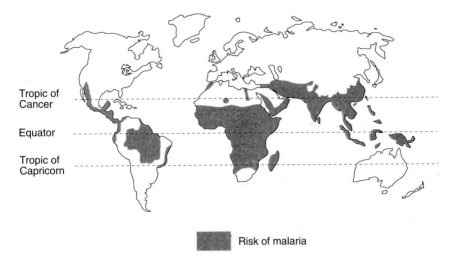

Risk of malaria

**Figure 1.1:** Geographical incidence of malaria.

year. Although the overall case:fatality ratio is relatively low, more than 75% of malaria deaths in Africa occur in children under 5 years old.

It is estimated that 300 million people worldwide are infected with hepatitis B virus (HBV). This blood-borne pathogen is transmitted directly by sexual contact and by contaminated hypodermic needles, blood transfusions and blood products. The virus can also cross the placenta to infect the fetus or it can be passed peri- or postnatally from mother to child. HBV virus infections that become chronic result in those infected becoming carriers of the disease. In Western Europe, North America and Australia 0.1% of all infections become chronic, but this figure increases to 15–20% in tropical countries. A 25–30% risk of cirrhosis or hepato-carcinoma in chronic carriers adds significantly to the case:fatality ratio of hepatitis B infection.

Respiratory and gastrointestinal communicable diseases are the major causes of death of young children, particularly in developing countries. Pertussis (whooping cough) is responsible for high mortality in children, although another bacterial respiratory disease, tuberculosis, accounts for a higher overall case:fatality ratio. Measles, a viral respiratory disease, exacts the greatest toll in childhood, accounting annually for 1.5 million deaths. Typhoid is a major bacterial cause of gastrointestinal disease in all age groups but diarrhea caused by various infectious and parasitic agents is primarily responsible for deaths of approximately 5 million young children each year.

A bacterial infection that accounts annually for 750 000 newborn deaths is neonatal tetanus. This disease is caused by nonsterile obstetric procedures causing contamination of the umbilical cord at the time of delivery. Also, in some areas of the developing world, it is the local custom to cover the umbilical stump with mud or dung. Consequently, there is a high risk of infection because tetanus spores are commonly found in the soil and in the gastrointestinal tract of humans and animals in all parts of the world.

The most recent infectious agent to make a global impact in terms of pandemic spread in both developed and developing countries is the human immunodeficiency virus (HIV). This virus is another blood-borne infection with routes of transmission similar to those of HBV. HIV infection was first discovered in 1983 to be the cause of AIDS. Within 5 years the number of AIDS cases in one of the world's most highly developed countries, the United States of America, accounted for 56% of the global sum. By 1990 a total of 179 countries or territories were reporting AIDS cases to the World Health Organization (WHO). At that time the cumulative number of AIDS cases was 900 000 in adults and 400 000 in children, but it is now estimated that at least 10 million people have been infected worldwide, although many remain asymptomatic. In several developed and developing countries AIDS has now become the major cause of death in young adults. Such frightening statistics should, however, be set against those for influenza, a viral respiratory disease. During this century there have been several influenza pandemics, the most severe in 1918–19 causing over 20 million deaths – more than the total number of fatal casualties in the First World War.

Although other communicable diseases may have low or negligible mortality rates, they can leave a tragic legacy in terms of permanent disability. It is estimated that there are between 27 and 35 million blind people in the world. Trachoma, a chlamydial infection, and onchocerciasis or river blindness, a parasitic disease, have together been responsible for over 30% of world blindness. Paralysis of the upper or lower limbs due to poliomyelitis is still a significant cause of disability in developing countries despite the virtual eradication of this acute viral disease from many developed countries. Diarrheal disease resulting in malnutrition can lead to retarded physical and mental development in survivors, who also show greater susceptibility to other diseases.

Communicable diseases also make a significant impact on the global economy. Despite a considerable wealth of natural resources, many developing countries cannot realize their true economic potential while they have high endemic levels of infectious and parasitic disease. In turn, developed countries must sustain the significant financial costs of protection against the importation of infectious diseases that are endemic in other parts of the world. For all these reasons, infectious and parasitic diseases need to be controlled and, ideally, eradicated.

## 1.2 A short history of vaccine development: Jenner's legacy

Preventative measures to limit the spread of communicable diseases have been practiced since ancient times. A simple, but often cruel, safeguard was to separate diseased from healthy individuals by isolation or quarantine but, occasionally, other effective measures were discovered. In 1854 a local epidemic of cholera in London, England, was stopped by closing a public pump that supplied water to the area affected by the disease. The water supply was found subsequently to have been contaminated with sewage. Since the beginning of the 20th century, before treatment with antibiotics or immunization became available, better housing conditions in England were responsible for a 50% reduction in the incidence of tuberculosis. However, adequate water supplies and accommodation can be provided only if the local economy can support these costs.

Communicable disease becomes particularly difficult to control if the pathogen is able to infect both human and animal hosts. This can enable re-introduction of disease into the human population from another animal reservoir. Such a cycle becomes even more complicated if a vector carries the disease between the animal and human hosts. In the 1970s, attempts to control malaria using the insecticide DDT (dichlorodiphenyl-trichloroethane) to eradicate the mosquito vector were finally abandoned because DDT-resistant mosquitos had appeared, and DDT was found to have toxic effects on other animals.

If exposure to an infectious or parasitic agent cannot be readily avoided, effective protective measures must be able to act after infection has taken place but before disease develops. How this may be achieved was first recognized by Edward Jenner who, in 1796, provided proof that "the inoculated Cow Pox proves a perfect security against the Small Pox". The word 'vaccine' is derived from *vacca* (Latin) meaning cow and the word 'vaccination' is now used to describe, in general terms, the process of immunization against disease.

Smallpox was a specifically human disease which, in the 18th century, was virtually pandemic. Between 1780 and 1800 in one city, London, England, smallpox killed over 36 000 people; the disease was responsible overall for nearly one out of every 10 deaths and nine out of 10 deaths in children under 5 years old. In several parts of Europe in the 18th century, however, it was known that an infection acquired from cows (cowpox) could protect against smallpox. This association was particularly noticeable in rural areas because milkmaids who contracted the mild cowpox infection did not catch smallpox. Consequently, they retained unblemished complexions while unprotected victims were disfigured by smallpox scars. Jenner is credited with the first scientific study of vaccination when he gave experimental inoculations of cowpox to several individuals who were shown subsequently to be resistant to smallpox.

In 1801 Jenner published a treatise on 'The Origin of the Vaccine Inoculation' which expressed his hope "that the annihilation of the Small Pox, the most dreadful scourge of the human species, must be the final result of this practice". From the start of the 19th century, vaccination against smallpox was quickly adopted in Europe and North America, and was made compulsory in England in 1853, although such legal obligation had first been introduced in Sweden in 1807 (*Figure 1.2*). Smallpox, however, remained a savage scourge of the human race for nearly two centuries.

It was 80 years after Jenner's seminal studies before other vaccines were developed. They arose as a direct result of the emergence in the late 19th century of the 'germ theory of human disease'. This was finally given credibility by Robert Koch in 1876 when he isolated the anthrax bacillus in pure culture in the laboratory and showed that the inoculated bacterium transmitted the disease to mice. By the end of the next decade there was an impressive list of pathogens identified with specific infectious diseases: the tubercle bacillus, cholera vibrio, typhoid bacillus, diphtheria bacillus and tetanus bacillus. The development of techniques for growth *in vitro* of the causal agents of infectious diseases also permitted a discovery of major importance. In 1888 Roux and Yersin showed that bacteria-free filtrates of diphtheria bacillus cultures could cause disease in laboratory animals. Within a short time other soluble bacterial exotoxins were discovered, notably those responsible for gas gangrene, botulism and tetanus. Some bacteria were found subsequently to have endotoxins (lipopolysaccharides associated with their outer membranes).

**Figure 1.2:** Smallpox vaccination clinic, Paris *c*. 1870.

Contemporaries of the pioneering bacteriologists were identifying the life histories of other human pathogens. In 1878 Patrick Manson first demonstrated mosquitos to be vectors of the filarial worm *Wuchereria bancrofti*, the cause of elephantiasis. He also showed there to be a necessary period of development of the parasite in the insect before it becomes able to transmit the infection. A decade later, Ronald Ross demonstrated the mosquito transmission of the malaria parasite.

The late 19th century saw the first determined attempts to apply newly acquired knowledge to the prevention of infectious diseases. The epigram, "Chance favors only the prepared mind", is credited to Louis Pasteur, and it applies *par excellence* to his development of vaccines. He found accidentally that chickens injected with an old culture of chicken cholera microorganism were resistant to subsequent challenge with a fresh, virulent culture. Over a period of 4 years Pasteur produced other vaccines. Wild-type rabies virus ('street virus') was passaged intracerebrally in rabbits to produce a modified or 'fixed' virus, and Pasteur's original rabies vaccine was prepared by drying spinal cords taken from rabbits infected with the fixed virus. An anthrax vaccine was produced by growing the bacterium at an elevated (42–43°C) temperature and, in 1881, Pasteur conducted a well-publicized field trial in which several sheep, cows and a goat were inoculated twice with this vaccine. Two weeks later the vaccinated animals and unvaccinated control animals were injected with a virulent culture of the anthrax bacillus. Within 2 days the unvaccinated animals had died but the vaccinated animals survived.

In 1890, von Behring and Kitasato immunized animals with the diphtheria toxin modified by chemical treatment with iodine trichloride such that its toxicity was inactivated. A year later this animal antiserum was used to treat a child with diphtheria. However, it was not until 1923 that Ramon introduced the formaldehyde-inactivated toxin or 'toxoid' human vaccine that is now used on a worldwide basis. In parallel with their work on diphtheria, von Behring and Kitasato also initiated development of the tetanus toxoid vaccine.

In 1923 another important bacterial vaccine was introduced in France, the tuberculosis vaccine bacille Calmette–Guérin (BCG). It is based on a bovine strain that was attenuated, that is its human pathogenicity was diminished, by repeated passage in culture in bile-containing media. The BCG vaccine was very controversial for many years because of doubts concerning its reversion to virulence but such fears have not been justified. To date, 3 billion people have been immunized with BCG although this vaccine has not been used extensively in some countries, including the United States of America.

*Bordetella pertussis*, the causative agent of whooping cough, was first isolated by Bordet and Gengou in 1906, but pertussis vaccines, based either on killed whole organisms or on crude extracts, did not enter general use until the late 1940s. There has been considerable controversy concerning the safety of such pertussis vaccines, particularly with regard to serious reactions such as encephalopathy. Because it acts as an adjuvant, that is it enhances the immunogenicity of other soluble proteins, pertussis vaccine is often given together with diphtheria and tetanus vaccines (DPT vaccine).

At the end of the 19th century, yellow fever was a scourge of American soldiers fighting in Cuba during the Spanish–American War. In 1900–1901 the American Army Commission under Major Walter Reed pursued the observations of a Cuban physician to show, by human volunteer experiments, the transmissible nature of the disease together with the involvement of the mosquito *Aedes aegypti* as an insect vector. Yellow fever was found to be a viral disease by Stokes, Bauer and Hudson in 1927 but initial control measures were based on eradication of the urban insect vector. However, this approach became untenable when, in the mid-1930s, a cycle of 'jungle yellow fever' with animal reservoirs was discovered involving monkeys and *Haemagogus* mosquitos in the New World, and monkeys and certain *Aedes* mosquitos in central Africa. Fortunately, at the same time, the fertile hen's egg had been shown to support the growth of some viruses and in 1937 the development by Max Thieler of the 17D strain of yellow fever vaccine was announced.

Other viral vaccines did not appear until the late 1940s when Enders and others developed the cell and tissue culture techniques that have subsequently enabled the growth *in vitro* of many human viruses. Poliovirus grown in monkey kidney cell culture and killed with formaldehyde was introduced in 1955 by Jonas Salk as the inactivated poliovirus vaccine (IPV) administered by intramuscular inoculation. In 1957 an alternative vaccine was developed by Albert Sabin who attenuated poliovirus by repeated passage through cell culture to obtain the live vaccine given by mouth – the oral poliovirus vaccine (OPV). In 1962 other attenuated virus vaccines were developed for mumps and measles. Following an extensive epidemic of German measles in the United States in 1964 that resulted in the birth of 20 000 children with congenital disabilities, an attenuated rubella viral vaccine was made.

In 1968 a tragic event occurred which should serve as a cautionary note in vaccine development. Respiratory syncytial virus (RSV) is an important agent of disease in children. It causes bronchiolitis almost exclusively in infants under 1 year of age. In 1962 the virus was grown successfully in cell culture for the first time, and shortly afterwards a formaldehyde-inactivated vaccine was made. Clinical trials were carried out with infants from 2 months to 7 years of age, and administration of this vaccine resulted in seroconversion, that is the induction of RSV-specific antibodies. Sadly, these children were not protected against disease. Infants in the youngest age group at the time of immunization were found to be at much greater risk of developing severe lower respiratory tract illness when they were infected naturally with RSV compared with nonvaccinated children of the same age. Two infants died who were 2 and 5 months old when vaccinated and 14 and 16 months old, respectively, when they became ill. Rather paradoxically, the risk of disease did not appear to be altered in the vaccinees but severe illness was not seen in the older vaccinees. Similar worsening of disease was also seen in children immunized with a formaldehyde-inactivated measles virus vaccine who developed severe reactions after natural exposure to the measles virus.

## 1.3 Vaccination practice in developed countries

Most of the vaccines used today in developed countries have been available for at least 30–40 years. They are mainly attenuated, inactivated or toxoid vaccines. However, a different rationale has been used to develop a vaccine to protect children against bacterial meningitis caused by *Haemophilus influenzae*. Very young children are in greatest danger of infection by this organism because effective immunity does not develop until 2–3 years of age. The bacterium possesses a polysaccharide capsule which is a polymer composed of alternating units of ribose and ribitol, joined covalently by phosphoric diester linkage and referred to as 'PRP'. It is a major virulence factor and the target for antibodies which protect against disease. In 1983 clinical trials of a so-called 'subunit vaccine' based on the polysaccharide capsule of *H. influenzae* serotype b (Hib) were carried out in Finland. It proved to be poorly immunogenic and, to try to overcome this problem, Hib vaccines have been developed with the capsular polysaccharide attached covalently to the diphtheria or tetanus toxoid.

Vaccines currently used in developed countries include:

| | |
|---|---|
| DPT | diphtheria, pertussis and tetanus |
| MMR | mumps, measles and rubella |
| BCG | tuberculosis |
| OPV or IPV | poliomyelitis |
| PRP | *H. influenzae* (Hib) meningitis |

In developed countries, following the introduction of new vaccines or vaccination programs, the incidence of many communicable diseases has declined markedly (*Table 1.3*).

## 1.4 Vaccination practice in developing countries

Since 1974 the WHO has promoted the Expanded Program on Immunization (EPI). It is directed mainly against infectious diseases that occur in infants and young children – diphtheria, measles and pertussis – but also those which occur in adults – tuberculosis and poliomyelitis. Antenatal vaccination of mothers is used to prevent neonatal tetanus. In certain parts of the world, EPI also includes other

**Table 1.3:** Reductions in morbidity and mortality due to infectious diseases in England and Wales following the introduction of new vaccines or vaccination programs

| Disease | Annual cases | Annual mortality |
|---|---|---|
| Diphtheria | | |
| 1939 | 47 343 | 2133 |
| 1940 | Vaccination program | |
| 1967 | 6 | 0 |
| Pertussis | | |
| 1951 | 169 347 | 453 |
| 1957 | Vaccination program | |
| 1989 | 11 646 | 2 |
| Polio | | |
| 1955 | 6331 | 270 |
| 1956 | Vaccine introduced | |
| 1970 | 6 | 0 |
| Measles | | |
| 1967 | 460 407 | 99 |
| 1968 | Vaccine introduced | |
| 1990 | 13 301 | 0 |
| Congenital rubella | | |
| 1970 | 60 | 0 |
| 1970 | Vaccination program | |
| 1990 | 1 | 0 |

vaccines to protect against other diseases with a restricted geographical distribution – yellow fever in parts of Africa and Japanese encephalitis, a killed virus vaccine, in the Orient (*Table 1.4*).

carboxy The hepatitis B vaccine is likely to be added to the WHO EPI program. New methods had to be developed to produce this vaccine because HBV cannot be grown reliably in cell cultures (see Chapter 10). In 1982 molecular biological techniques were used for the first time to produce a recombinant hepatitis B vaccine. It is based on the knowledge that protective immunity against HBV infection is induced by a glycoprotein of the virion, the HBV surface antigen (HBsAg). The virus genomic DNA sequence that encodes this antigen was cloned and expressed using a yeast vector. When introduced in 1985, the hepatitis B

**Table 1.4:** Infectious diseases in the WHO EPI

Tuberculosis
Diphtheria
Pertussis
Tetanus
Poliomyelitis
Measles
Yellow fever in parts of Africa
Japanese encephalitis in parts of the Far East

vaccine was the first vaccine generated by recombinant DNA technology to be licensed for human use.

## 1.5 The need for new and improved vaccines

*Table 1.5* includes a list of vaccines that are still required or need to be improved to prevent communicable diseases.

In developing countries, over half of the deaths in young children are caused by respiratory infections with *Streptococcus pneumoniae, H. influenzae, Staphylococcus aureus*, influenza virus, RSV and parainfluenza virus. *S. pneumoniae* and *H. influenzae* also cause meningitis although *Neisseria meningitidis* is another important etiological agent of this disease. Diarrhea, the second major cause of death due to infectious disease, is caused by a multiplicity of etiological agents including several bacteria belonging to the genera *Enterobacteriaceae* with viral diarrheas caused mainly by rotaviruses. Enteric diseases can also be caused by parasitic protozoa and helminth infections.

Sexually transmitted diseases (STD) are a worldwide problem. They create considerable physical and psychological trauma in infected individuals, together with a considerable economic burden. The HIV pandemic, because of its high case:fatality ratio, is currently causing great concern but *Chlamydia trachomatis* is the most common STD; gonorrhea, syphilis and genital herpes are endemic in many developed and developing countries with *Haemophilus ducreyi* infections being more common in the subtropics.

Cancer is a feared disease with multifactorial causes. Amongst infectious agents Epstein–Barr virus (EBV) has been shown to be causally associated with certain human cancers: Burkitt's lymphoma and nasopharyngeal carcinoma (NPC). A strong case can be made for the development of vaccines to control NPC and other EBV-associated tumors. HBV infection is linked with subsequent development of hepatocarcinoma. Although hepatitis B vaccines are already available, improvements need to be made.

Of all parasitic diseases, malaria is the major cause of morbidity and mortality on a global basis. However, other insect vector-borne protozoa

**Table 1.5:** Diseases or pathogens requiring new or improved human vaccines[a]

*Chlamydias*, cholera[b], coronaviruses, cytomegalovirus, dengue, Epstein–Barr virus, *E. coli*, filariasis, *G. lamblia, H. ducreyi*, hepatitis A, B[b] and C, herpes simplex type 2, HIV-1 and -2, influenza[b], leishmaniasis, malaria, *Meningococcus* A/C[b] and B, *Mycobacterium leprae, Mycobacterium tuberculosis, Mycoplasma pneumoniae, Neisseria gonorrhoea*, parainfluenza, pertussis[b], poliovirus, respiratory syncytial virus, rotavirus, *Salmonella typhi*[b] schistosomiasis, *Shigella, S. pneumoniae*[b] *S. aureus*, treponema, trypanosomiasis, yellow fever[b]

[a]Based on ref. [2].
[b]Improved.

are the flagellated trypanosomes responsible for sleeping sickness in equatorial Africa and Chagas' disease in South America. *Leishmania* are protozoan parasites that cause visceral disease or skin lesions in Central and South America, the Middle East and India. Another protozoan, *Giardia lamblia*, is a water-borne pathogen responsible for high morbidity and long-term effects due to malnutrition in both developed and developing countries. Helminth or worm infections are severe, debilitating diseases that include filariasis. This is caused by mosquito-transmitted nematodes that invade the lymphatics leading to gross enlargement known as 'elephantiasis'. To date, there are no vaccines in general use to control human parasitic diseases.

Although some vaccines have been used for a number of years, doubts remain concerning their safety or efficacy. The incidence of whooping cough has been reduced markedly following the introduction, over 40 years ago, of a vaccine comprising a suspension of killed *B. pertussis* organisms. However, there is still significant controversy concerning its side effects: crying, screaming and fever may occur, but severe neurological complications resulting in brain damage and death have also been reported after children have been given pertussis vaccine. Reversion to virulence by the live, attenuated poliovaccine (OPV) is responsible for poliomyelitis at a frequency of about one to two cases per million vaccinations. Improvements are clearly necessary if global eradication of poliomyelitis is to be attempted. Influenza vaccines need to be reviewed regularly because the viruses in circulation change unpredictably due to the genetic lability of this virus. Killed virus vaccines, the most widely used influenza vaccines, confer total protection on only about 30% of vaccinees although about 70% have reduced disease following exposure to influenza.

To design safe and effective vaccines it is necessary to have detailed knowledge of the relevant pathogen and how it causes disease. Clearly, the host's ability to mount immune responses must also be understood and placed in the context of the immunity that is actually induced in response to infectious and parasitic agents. Then it may be possible to determine how to elicit appropriate protective immune responses and to use genetic engineering and other modern techniques to design suitable vaccines.

## References

1. WHO (1993) *World Health Statistics Annual (1990–1993)*. World Health Organization, Geneva.
2. Robbins, A. (1990) *Lancet,* **335,** 1436.

Chapter 2

# Etiological agents of communicable diseases requiring new or improved human vaccines

A brief description follows of the etiological (causal) agents of infectious and parasitic diseases targeted by the WHO EPI (see *Table 1.4*), together with those of communicable diseases for which new or improved vaccines are still required (see *Table 1.5*).

The relative sizes of these pathogens are shown in *Figure 2.1*. They range from viruses with dimensions measured in molecular terms through to worms which can reach a meter or more in length.

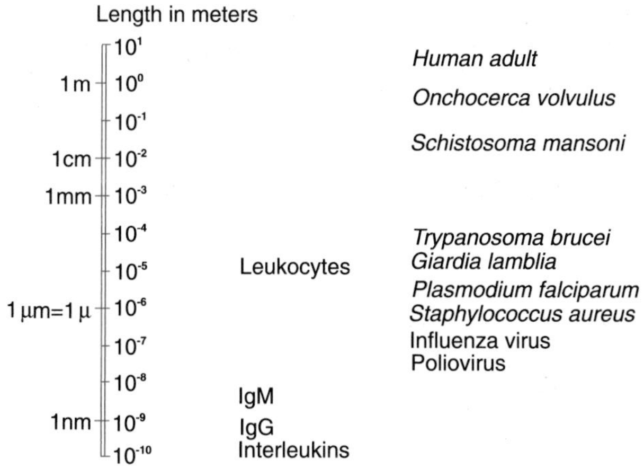

```
Length in meters
         ┌ 10¹
1m  ─┤  10⁰            Human adult
         ├ 10⁻¹          Onchocerca volvulus
1cm ─┼ 10⁻²          Schistosoma mansoni
1mm ─┼ 10⁻³
         ├ 10⁻⁴          Trypanosoma brucei
         ├ 10⁻⁵   Leukocytes   Giardia lamblia
                                       Plasmodium falciparum
1μm=1μ ─┼ 10⁻⁶          Staphylococcus aureus
         ├ 10⁻⁷          Influenza virus
                                       Poliovirus
         ├ 10⁻⁸   IgM
1nm ─┼ 10⁻⁹   IgG
         └ 10⁻¹⁰  Interleukins
```

**Figure 2.1:** Relative sizes of molecules and cellular components of host innate and acquired immune defence mechanisms, various pathogens and their human host.

## 2.1 Viruses

Viruses (see *Table 2.1*) are obligate intracellular parasites, that is they can multiply only in living cells and they cannot be grown on artificial culture media. The propagation of viruses in the laboratory can be carried out only in tissue culture, fertile hen's eggs or laboratory animals. Cell culture systems are particularly useful to study the molecular events associated with virus growth. However, the marked species specificity of certain human viruses has severely limited the study of their pathogenic mechanisms and induced immune responses. For example, the lack of an alternative animal host for HIV has severely restricted the development of AIDS vaccines.

Virus particles or virions are too small (0.02–0.35 µm) to be seen by ordinary light microscopy and are discernible only by electron microscopy. The viral genome is either RNA or DNA; viruses are unusual in not containing both types of nucleic acid and, in some cases, using RNA as a genome. The genome is surrounded by a protein coat (capsid) arranged symmetrically in a regular pattern of subunits with either icosahedral or helical symmetry. Some virions also have an envelope derived from modified host cell membranes (*Figure 2.2*). Nonenveloped virions, therefore, are composed chemically of nucleic acid and protein only, although phospholipid and carbohydrate are also found in enveloped viruses. Virions do not contain subcellular organelles such as the mitochondria, ribosomes or Golgi apparatus found in either prokaryotic or eukaryotic cells, and the extracellular virion is metabolically inert.

**Table 2.1:** Diseases and their associated viruses

| Disease | Etiological agent(s) |
| --- | --- |
| AIDS | Human immunodeficiency virus |
| Bronchiolitis | Parainfluenza viruses |
|  | Respiratory syncytial virus |
| Cancer | Epstein–Barr virus |
| Congenital defects | Rubella virus |
|  | Human cytomegalovirus |
| Croup | Parainfluenza viruses |
| Dengue | Dengue virus |
| Gastroenteritis | Rotaviruses |
|  | Caliciviruses |
|  | Coronaviruses |
| Genital herpes | Herpes simplex virus (HSV) type 2 |
| Hepatitis | Hepatitis A, B and C viruses |
| Influenza | Influenza viruses |
| Japanese encephalitis | Japanese encephalitis virus |
| Measles | Measles virus |
| Mumps | Mumps virus |
| Pneumonia | Parainfluenza viruses |
|  | Respiratory syncytial virus |
| Rubella (German measles) | Rubella virus |
| Poliomyelitis | Polioviruses |
| Yellow fever | Yellow fever virus |

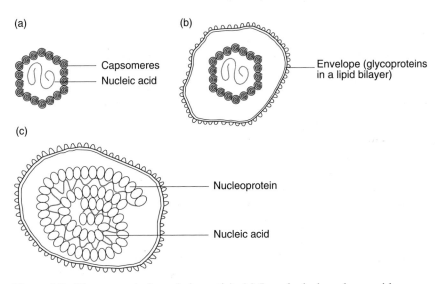

**Figure 2.2:** Virus morphology (schematic). (a) Icosahedral nucleocapsid, nonenveloped. (b) Enveloped icosahedral nucleocapsid. (c) Enveloped helical nucleocapsid.

A productive virus replication cycle consists of the following sequential events: (1) attachment to the host cell surface; (2) penetration into the cytoplasm; (3) release of the viral genome within the infected cell; (4) replication of viral nucleic acid and proteins; (5) assembly of new virus particles; and (6) their release from infected cells which may be accompanied by the acquisition of an envelope (*Figure 2.3*).

Initial attachment of a virion to the host cell membrane is sometimes made via a cellular receptor. This is increasingly being seen as an explanation for the tissue tropism exhibited by viral pathogens. For example, HIV virions attach via their envelope glycoprotein gp120 to the CD4 molecule found on the surface of a particular subset of T lymphocytes, and the gp340 envelope glycoprotein of EBV attaches to a complement receptor, CR2, on B lymphocytes. Once adhered to the surface, entry of a virion into its host cell is effected either by phago-cytosis or by fusion of the virion envelope with the cell membrane followed by endocytosis. This is followed by removal of the capsid (uncoating) within the cytoplasm to release the viral nucleic acid into the infected cell.

Viral nucleic acid replication takes place either in the cytoplasm (mainly RNA viruses) or nucleus (some RNA viruses but all DNA viruses except poxviruses, e.g. vaccinia virus). Replication of viral nucleic acid clearly requires virus-encoded polymerases in the case of all RNA viruses. The RNA genome of one important virus family, the retroviruses, has a unique mechanism of replication: a virus-encoded enzyme, reverse transcriptase, utilizes the viral RNA genome as a template to produce a

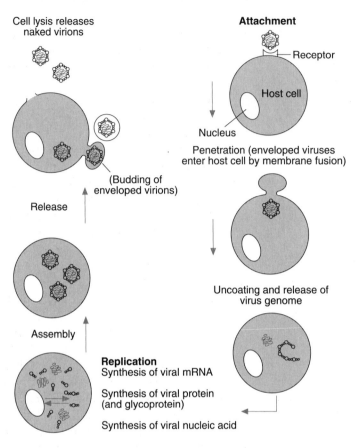

**Figure 2.3:** The virus replication cycle (schematic).

complementary DNA copy (proviral DNA) which is integrated into host cell chromosomal DNA. This is an essential step in the replication of an important human retrovirus, HIV. Small DNA viruses utilize host cell enzymes to replicate but the larger herpesviruses do encode a virus-specific DNA polymerase.

Of the viruses listed in *Table 2.1*, positive-sense single-stranded viral RNA genomes (poliovirus) can function directly as messenger RNA (mRNA) molecules immediately after uncoating has occurred. It is a peculiar feature of other RNA viruses with negative-sense genomes (influenza viruses) or with double-stranded RNA genomes (rotaviruses) that a transcriptase carried by the virion produces a positive-sense copy. This is processed further, by polyadenylation and methylation using cellular, nuclear enzymes, to make functional viral mRNA. Transcription of viral DNA, including the integrated proviral DNA of retroviruses, is carried out by host cell enzymes with one exception: poxviruses (vaccinia virus) are obliged to carry full transcriptional machinery due to their intracytoplasmic site of replication. Translation of virus-specific mRNA molecules takes place on host ribosomes, often to the detriment of host-specific protein synthesis which may consequently be inhibited completely. In addition to native molecules, glycosylated and other post-translationally modified virus-specific proteins may be produced.

The assembly of viral nucleic acid and protein to produce progeny virions takes place within either the cytoplasm or the nucleus of the host cell. The mechanisms that release newly formed virions are related to their morphogenesis. Some virus particles consist solely of a nucleocapsid, and they are often released simply by lysis of the infected cell (poliovirus). However, some viruses have an envelope which is derived from host cell membranes that have been modified by insertion of virus-specific glycoproteins. These virion envelopes are derived either from the nuclear membrane (herpesviruses) or from the plasma membrane (influenza viruses and HIV). Consequently, the host cell may remain alive for some time and continue to shed enveloped virus particles.

Viruses have various cytopathic effects on the infected host cell. Polioviruses are cytocidal, that is completion of the replication cycle is followed rapidly by death and lysis of the infected cell. Other viruses, particularly the enveloped viruses, must maintain the integrity of the infected cell at least while the new virions bud from its surface. Other enveloped viruses, HIV and RSV, cause cell fusion. Persistent or chronic infections, for example by hepatitis viruses and HIV, enable the prolonged production of progeny virions. Herpesviruses have the unique ability to establish latent infections in which they can remain quiescent for long periods after infection before they are eventually reactivated to complete a productive replication cycle resulting in cell death. Finally, certain viruses, such as EBV, can 'immortalize' cells. Some infected cells are driven into cycles of mitosis, possibly by the production of virus-specific factors which stimulate cell division. Such virus-infected host cells in a natural EBV infection may undergo a malignant transformation into cancer cells.

## 2.2 Bacteria

Many bacteria are free-living and may be cultured readily in the laboratory. Discrete bacterial colonies can be isolated on solid media, and selective media may be used to recover particular bacteria from a mixed population. Oxygen requirements differ considerably: for example, there are obligate aerobes (*M. tuberculosis*), obligate anaerobes (*Clostridium tetani*) and facultative organisms (*E. coli*) able to respire either aerobically or anaerobically. Certain bacteria, mainly aerobic bacilli and anaerobic clostridia, can form highly resistant spores and remain dormant for long periods.

Bacteria are unicellular, prokaryotic organisms which, unlike eukaryotic cells, do not have separate nucleus and cytoplasm, but a single, circular chromosome and no nuclear membrane (*Figure 2.4*). There are no mito-chondria and no endoplasmic reticulum or Golgi apparatus. Bacterial ribosomes have a sedimentation coefficient of 70S compared to the mammalian 80S ribosomes, and initiating transfer RNA is formylmethio-nyl RNA rather than eukaryotic *N*-methionyl tRNA. Multiplication of bacteria occurs by binary fission.

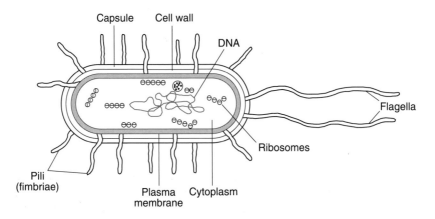

**Figure 2.4:** Bacterial structure (schematic).

Most bacteria have a size of about 1 μm ($10^{-6}$ m). A rigid cell wall gives typical morphological appearances which are mainly spherical cocci or rod-shaped bacilli, although vibrios are curved. Bacteria can be distinguished cytochemically by the Gram stain: Gram-positive organisms retain the initial stain (crystal violet fixed with iodine) following washing with an organic solvent (acetone or ethyl alcohol) whereas Gram-negative bacteria are decolorized and are visualized by counter-staining with a different color (saffranin). Hence, the bacteria listed in *Table 2.2* may be categorized readily: Gram-positive cocci (staphylococci, streptococci); Gram-negative cocci (*Neisseria*); Gram-positive rods (bacilli, clostridia); and Gram-negative rods (*E. coli, Salmonella, Shigella*). Tubercle bacilli and other *Mycobacteria* are 'acid-fast': they retain color after staining with hot carbol fuchsin and washing with acid alcohol. Mycobacterial cell walls have a high (60%) lipid content which includes mycolic acids (large, saturated α-alkyl, β-hydroxyl fatty acids).

Some bacteria (*S. typhi*) are motile because they possess flagella, and most Gram-negative bacteria have fimbriae or pili which are surface appendages used for attachment (*Figure 2.4*). The surface of some bacteria is covered by a capsule or slime layer; most capsules are polysaccharides (streptococci) but some are polypeptides. The rigid bacterial cell wall has unique components such as muramic acid, D-amino acids and diaminopimelic acid (*Figure 2.5*). Gram-positive bacterial cell walls also contain teichoic acids and Gram-negative cell walls are covered with lipopolysaccharide (LPS).

**Table 2.2:** Diseases and their associated bacteria

| Disease | Etiological agent(s) |
|---|---|
| Bacillary dysentery | *Shigella* spp. |
| Chancroid | *H. ducrei* |
| Cholera | *Vibrio cholerae* |
| Diphtheria | *Corynebacterium diphtheriae* |
| Gastroenteritis | *E. coli* |
| Leprosy | *Mycobacterium leprae* |
| Meningitis | *H. influenzae* |
|  | *N. meningitidis* |
| Syphilis | *Treponema pallidum* |
| Tetanus | *Clostridium tetani* |
| Tuberculosis | *M. tuberculosis* |
| Typhoid | *S. typhi* |
| Whooping cough | *B. pertussis* |
| Various | *Streptococci* |

Bacterial pathogenicity is often mediated by toxin production. Exotoxins are proteins secreted by growing Gram-positive or Gram-negative bacteria causing diseases such as diphtheria and tetanus. Such bacterial products are amongst the most powerful poisons known; 1 ng ($10^{-9}$ g) of the tetanus exotoxin will kill a guinea pig. The precise effects of many bacterial exotoxins are due to their specific mechanisms of action. For example, the spastic paralysis caused by tetanus is due to the toxin's action as an inhibitor of the release of acetylcholine from the motor end plate thus preventing the transmission of signals from nerve to muscle. Endotoxins are the LPS components of the cell walls of Gram-negative bacteria released autolytically from dead bacteria. They are less toxic than exotoxins and less specific in their action. Diseases mediated by endotoxins often result in fever and septic shock.

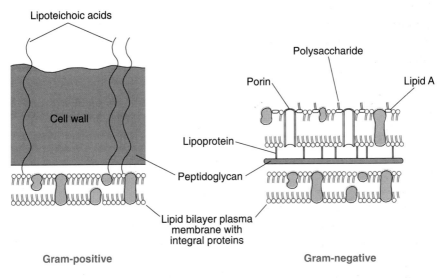

**Figure 2.5:** Bacterial cell walls.

## 2.3 Chlamydia

Unlike other bacteria, chlamydia (see *Table 2.3*) are obligate intracellular organisms. They have a rather complex life cycle consisting of two morphological forms, one adapted to intracellular multiplication and the second to extracellular survival (*Figure 2.6*). The infectious form of the organism, the elementary body, is very small (diameter 0.3 µm) with a double-layered outer membrane but no cell wall murein layer. In its extracellular form, the elementary body attaches to a susceptible host cell and enters by receptor-mediated endocytosis. Multiplication occurs by binary fission in the cytoplasm of the host cell with formation of larger (0.5–1.0 µm) reticulate bodies, so-called for their mesh-like appearance in stained preparations. Reticulate bodies are fragile and revert to elementary bodies before progeny leave the cell. They may be released by cell disruption although others survive by extrusion of the intact endosome.

**Table 2.3:** Diseases and their associated chlamydia.

| Disease | Etiological agent |
| --- | --- |
| Trachoma<br>Lymphogranuloma venereum | } *C. trachomatis* |

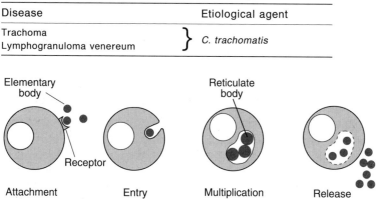

**Figure 2.6:** The life cycle of *Chlamydia*.

## 2.4 Parasitic protozoa

Unlike acute disease caused by many bacterial and viral infections, parasitic diseases are frequently chronic. Parasitic infections may involve more than one host and transmission to a human host may be accidental. For some parasites, however, the human host is obligatory although other necessary stages of its development take place in other animals.

Parasitic protozoa (see *Table 2.4*) are unicellular, eukaryotic organisms (*Figure 2.7*). They may be either intracellular or extracellular within the infected host, and often exist in two forms. The active trophozoite is associated with growth and multiplication, and the dormant cyst is resistant to the various physical and chemical hazards it encounters during transition from one host to another. Intracellular parasites are typically transmitted between human hosts by insect vectors. Extracellular parasites include intestinal forms which are maintained in a human population through the fecal–oral route.

**Table 2.4:** Diseases and their associated parasitic protozoa

| Disease | Etiological agent |
|---------|-------------------|
| Diarrhea | *G. lamblia* |
| Leishmaniasis | *Leishmania* spp. |
| Malaria | *Plasmodium* spp. |
| Trypanosomiasis | *Trypanosoma* spp. |

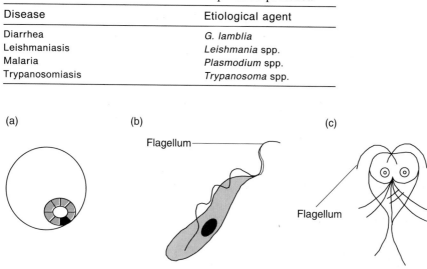

**Figure 2.7:** Protozoan parasites. (a) *Plasmodium falciparum* trophozoite in red blood cell. (b) *Trypanosome brucei* trypanomastigote. (c) *G. lamblia* trophozoite.

The malaria parasite requires both vertebrate and insect hosts. It has a complex life cycle with four phases of development, each with a distinctive morphology (*Figure 2.8*). The sporozoite, a slender, fusiform organism, is injected directly into the bloodstream of a vertebrate host by feeding mosquitos and enters the liver. As a fixed tissue form it matures to become a merozoite that can invade red blood cells. Here there are several cycles of schizogony (asexual reproduction) before cell lysis occurs and more merozoites re-enter the bloodstream to infect other erythrocytes. Some merozoites become sexual gametocytes which a feeding mosquito withdraws in its blood meal; gametogenesis then takes place in the gut lumen of the insect. The male gamete fuses with the female gamete to produce a zygote which develops into an oocyst and attaches to the insect's stomach epithelium. Sporogony occurs to produce sporozoites which enter the salivary gland to be injected into the vertebrate host in saliva by the feeding insect and so complete the cycle.

*Leishmania* spp. are protozoa transmitted by sandflies of the *Phlebotomus* genus, found generally in tropical or subtropical areas but rarely in temperate regions. The promastigote, an elongated organism with a flagellum (which resembles trypanosomes), is injected into the skin by a feeding insect. The parasite attaches to macrophages by its flagellum and becomes internalized, when transformation into the round, non-flagellated amastigote occurs. Repeated cycles of asexual reproduction occur with the infected cells remaining either in the skin (cutaneous leishmaniasis) or traveling to deeper organs (visceral leishmaniasis, kala-azar). Amastigotes in the bloodstream are ingested by feeding insects

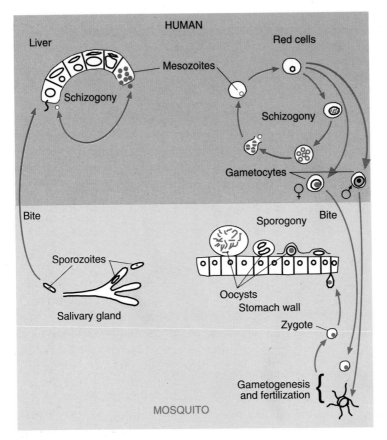

**Figure 2.8:** The life cycle of *Plasmodium vivax*.

where they transform into the flagellated promastigotes which multiply in the sandfly's gastrointestinal tract.

The two main trypanosomal diseases of man are sleeping sickness caused by *Trypanosoma brucei* in tropical Africa, and Chagas' disease caused by *T. cruzi* in Central and South America. Again, the parasite is a flagellated protozoon with a life cycle involving an intermediate host that is usually an insect. The vector for *T. brucei* is the tsetse fly, a member of the genus *Glossina*. The protozoan parasite has morphologically and metabolically distinguishable 'slender' or 'stumpy' bloodstream forms and tsetse fly midgut forms. Both forms enter the bloodsteam of the human host from salivary glands of a feeding insect, multiply in the blood and lymphatics to appear eventually in the central nervous system (CNS) and infect the brain.

*T. cruzi* is another flagellated protozoan which is transmitted by brightly colored bugs of the family *Reduviidae*. It multiplies in the midgut of the vector as an epimastigote and trypanomastigotes are voided in feces. Infection occurs when the insect bite-wound contaminated by feces is rubbed; eyelids and conjunctiva are common sites of infection. In the human host the trypano-mastigote transforms to an anamastigote and multiplies in the reticuloendothelial system. At this stage the parasite disappears from the blood for several days before more trypanosomes escape from ruptured cells to be taken to the peripheral blood vessels where the cycle is completed by further infection of reduviid bugs. In Chagas' disease caused by *T. cruzi* there are swellings and inflammation commonly around the eye ('chagomas'). More seriously, some infected persons develop severe neurological disorders many years later.

*G. lamblia* is an intestinal protozoan that has perfect bilateral symmetry (15 μm long, 9 μm wide and 3 μm thick) with a 'tear-drop' shape. There are four pairs of flagellae and on the ventral surface is an adhesive disk. It attaches to intestinal mucosa in the duodenum and multiplies asexually. In severe infections the trophozoites may cover large areas. Trophozoites are killed by drying in an external environment, and a resistant cyst form is responsible for fecal–oral transmission.

## 2.5 Parasitic helminths

Helminths (see *Table 2.5*) are roundworms (nematodes), tapeworms (cestodes) and flukes (trematodes). They are metazoa or multicellular organisms and the largest agents of infectious disease. This means they are extracellular and often have specialized attachment mechanisms to prevent their removal in a dynamic environment such as the alimentary canal or blood vessel within the infected host. Most helminths infect the intestinal tract although others penetrate deeper to internal organs. Cysts, the resting form, may be found within tissues. The life history of most helminths is complex and involves other animal reservoirs, although inanimate sources of infection may also occur. Transmission may be by insect vectors or the fecal–oral route, while certain worms have forms that are able to penetrate directly through human skin.

*W. bancrofti* was the first human blood parasite shown, by Manson in 1878, to be transmitted by an arthropod. It causes the fleshy deformities known as 'elephantiasis'. These filaria are long, thread-like nematodes transmitted by blood-feeding mosquitos (*Culex pipens*, also *Aedes* and *Anopheles* spp.). The sexually mature females (6.5–10 cm long and 0.2–0.28 mm in diameter) lie inextricably coiled in lymph glands or ducts, and they release microfilariae (180–300 μm long and 3–9 μm in diameter) in swarms. This shows periodicity, with microfilariae in blood at night but concentrated in the lungs during daytime. After ingestion by the biting mosquito host, further growth and differentiation takes place. Infective larvae are transmitted to human hosts by a bite from an infected mosquito and maturation to a sexually mature worm takes about 9 months.

*Onchocercus volvulus* is another filarial infection which causes serious dermatitis and 'river blindness'. Damage to the eye includes the cornea,

**Table 2.5:** Diseases and the associated parasitic helminths

| Disease | Etiological agent |
| --- | --- |
| Filariasis | Nematoda: *Filariina* |
| Schistosomiasis | *Schistosoma* spp. |

iris, conjunctiva and optic nerve, although the lesions may take a long time to develop – up to 40 years. Intermediate hosts are various species of blackflies of the genus *Simulium*. The developing worms become entangled in fibrous tissue as a result of a host inflammatory reaction. Adult worms may survive for many years in the lymphatics.

*Schistosomatidae* are trematodes found in the bloodstream, usually the mesenteric vein but also the venous plexus of the bladder. They are dioecious, that is different individuals have either male or female sex organs. The male (up to 75 mm long) bears the smaller female in a ventral canal, and one pair of worms can produce 100–300 eggs daily (*Figure 2.9*). The eggs are barbed to attach to blood vessel walls. Eventually, the burden of eggs causes the blood vessel to rupture, allowing eggs to reach intestinal lumen or bladder. After their release in feces or urine, the eggs hatch if exposed to water and light to release miracidia which enter snails by penetration of the shell. Here the miracidia mature to become sporocysts from which cercariae emerge to be released from the snail into surrounding water. The cercariae attach to the epidermis of a human host, penetrate the skin and enter venous circulation either directly or via the lymphatic system where they lose their tails and mature.

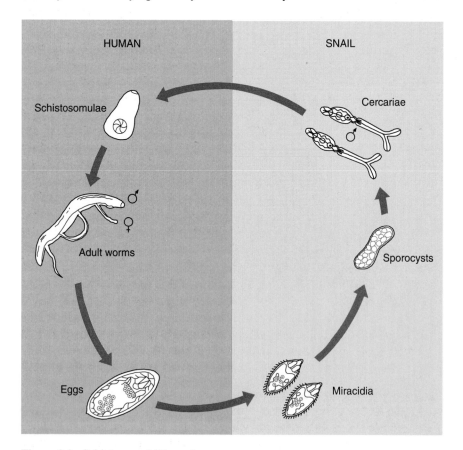

**Figure 2.9:** Schistosomal life cycle.

# Chapter 3

# Mechanisms of pathogenesis

Pathogenic organisms are able to cause disease, and the magnitude of this capability may be termed 'virulence'. This is determined mainly by their ability to invade and to cause damage to the host's tissues and organs. Such events are more likely to occur if the infectious agent can multiply rapidly. For example, one tissue-culture dose of infectious rhinovirus can cause a common cold, and 10 *Shigella dysenteriae* bacteria or 10 rotavirus particles can cause enteric disease, whereas 10 million *Salmonella* bacteria may be required to cause food poisoning. Other virulence factors of a pathogen include the ability to adhere to epithelial surfaces, to penetrate the body and reach selected target organs, and to produce soluble virulence factors such as toxins. However, the host's susceptibility to infection may be determined by its genetic constitution. A study of tuberculosis in human twins has shown both twins to be affected in 87% of identical twins but in only 26% of nonidentical twins. Such susceptibility appears to be determined by the ability to mount an immune response. Thus, pathogenicity reflects both quantitative and qualitative interactions between an infectious agent and its host.

Vaccines are designed to use immune intervention to tip the balance between infection and disease. This can be done most effectively only if the following questions are answered: (1) how does the pathogen gain entry into its host?; (2) how is localized infection established in the host?; (3) are toxins produced at the localized site of infection?; (5) does the pathogen spread systemically throughout the host?; (6) what target organs are attacked by the pathogen?

## 3.1 Anatomical sites used by pathogens to gain entry into the human body

Biological maintenance of the human body requires vital functions such as respiration, nutrition and excretion. These physiological activities are served by specialized structures: the nose, mouth, oropharynx, trachea and lungs for respiration; the mouth, esophagus and gastrointestinal tract for nutrition; and the urinary system and bowels for excretion. Because they are in close proximity to the external environment, these specialized anatomical areas readily serve as 'portals of entry' for pathogens into the human body. Deeper penetration into the host may be achieved more readily from such sites because they are lined by living epithelial tissues designed for the transport of solids, liquids or gases. A number of pathogens can also be transmitted sexually either as a result of lesions upon or within the sexual organs or through infected semen, vaginal secretions or blood.

Again, the pathogens described are those referred to in *Tables 1.4* and *1.5*, which are relevant to the WHO EPI or cause communicable diseases requiring new or improved vaccines.

### 3.1.1 Skin

Pathogens gaining access through the skin:

> *C. tetani* (localized infections)
> Dengue virus, hepatitis B and C viruses, HIV, Japanese encephalitis virus (JEV), rabies virus, yellow fever virus
> *Leishmania* spp.
> Nematoda: *Filariina*, *Plasmodium* spp., *Schistosoma* spp., *Trypanosoma* spp.

Unbroken skin is a good perimeter defense against infection because it presents a relatively impervious, mechanically resilient barrier to the outside world. The outermost epidermal layer consists of dead keratinocytes and provides particularly effective protection against pathogens such as viruses and other obligate intracellular parasites which require living cells for their survival. Consequently, pathogens are usually able to enter a host through the skin only if it has been damaged or penetrated physically.

*C. tetani* spores in soil enter deep wounds caused by contaminated sharp objects. Anaerobic conditions must exist in the infected wound for tetanus spores to germinate producing vegetative bacteria that release tetanospasmin. This toxin binds to ganglioside receptors and is transported to the CNS where it acts specifically against inhibitory synapses. This results in over-excitation of motor neurones with resultant increased muscle tone, rigidity (lockjaw) and spasms.

Animal or insect bites introduce infectious agents through the skin either mechanically or as an essential part of a parasite's life cycle. Mosquitos transmit yellow fever, dengue, Japanese encephalitis, malaria and filariasis; *Leishmania* is transmitted by sandflies and trypanosomes by sandflies and other blood-feeding insects. The cercariae of schistosomes are released into water and directly penetrate the skin by a combination of secreted enzymes and muscular contraction.

In addition to these natural routes, other blood-borne infections such as HIV or hepatitis B and C viruses (HCV) may also be introduced through the skin by blood transfusion, by needle-stick injury or by intravenous drug abuse.

### 3.1.2 Respiratory tract

Pathogens gaining access through the respiratory tract:

> *B. pertussis* (localized infections), *C. diphtheriae* (localized infections),
> *H. influenzae M. tuberculosis, M. leprae, N. meningitidis*
> EBV, influenza viruses (localized infections), measles virus, mumps
> virus, parainfluenza viruses (localized infections), RSV, rubella virus.

Nasal hairs and turbinates act as coarse filters to remove large particles
from inhaled air. Ciliated and goblet secretory cells in the respiratory tract
generate a mucous blanket that is continually swept upwards to the back
of the throat. This provides further protection apart from small particles
(less than 5 µm diameter) which are more likely to reach the alveoli.

The dynamic nature of the 'mucociliary escalator' requires a
respiratory pathogen to attach or adhere to the epithelium if it is to
establish an infection. The attachment of *C. diphtheriae* to mucosal
surfaces in the upper respiratory tract is followed by the production of
both neuraminidase (NA) and neuraminidate lyase. These enzymes break
down *N*-acetyl-neuraminic acid (NANA) residues to pyruvate and *N*-
acetylmannosamine to provide energy sources for corynebacterial growth.
Only strains infected lysogenically with a bacteriophage produce the
diphtheria exotoxin which is encoded by the bacterial virus. The toxin
binds to a membrane receptor, enters the epithelial cell and inhibits
protein synthesis. The resultant cell death causes an inflammatory
reaction producing the characteristic false membrane which can cause
mechanical obstruction of the larynx and trachea. Although infection
remains localized within the nasopharynx, systemic spread of the toxin
affects heart muscle leading to death due to heart failure.

*B. pertussis* adheres to bronchial ciliated epithelial cells where the
pertussis toxin appears to reduce migratory and phagocytic capabilities
together with the chemotactic responses of phagocytic cells. Such toxic
effects contribute to the prevalence of secondary bacterial infections.
Pertussis toxin also increases sensitivity to histamine and anaphylaxis.

*H. influenzae* attaches via pili causing acute local infection of the
pharynx and epiglottis. This can result in obstruction of the larynx which
may be fatal unless a tracheotomy is carried out. Intimate contact of *N.
meningitidis* with nasopharyngeal mucosal cells results in its engulfment
followed by transport to the submucosal space. Pathogenic streptococci in
the throat cause pharyngitis but systemic release of toxins also results in
scarlet fever, rheumatic fever and acute glomerulonephritis.

Influenza virus attaches to the epithelium using a viral envelope
glycoprotein, the hemagglutinin (HA), to bind to sialic acid residues on
epithelial cell surfaces. Progressive destruction of ciliated epithelium by
the cytolytic consequences of influenza virus replication leads to breakdown of the mucociliary escalator with resultant increased susceptibility
to secondary bacterial infection. Parainfluenza viruses, an important cause of

bronchiolitis and pneumonia in infants, also have envelope HA glycoproteins for attachment to host cells. In addition, there is a virion fusion protein (F) which enables infected cells to form syncytia, thus facilitating virus spread without exposure to the extracellular environment. The mumps virus envelope HA and F glycoproteins have similar functions. RSV, which causes bronchiolitis and pneumonia, is the most important respiratory pathogen during the first 2 years of life. This virus does not have a HA, and attaches to its host cell via another glycoprotein in the viral envelope but it does possess a fusion protein and forms syncytia.

### 3.1.3 Alimentary tract

*E. coli* (localized infections), *S. typhi, Shigella* spp., *V. cholerae* (localized infections).
Caliciviruses, coronaviruses, hepatitis A virus, polioviruses (localized infections), rotaviruses.
*G. lamblia* (localized infections).

Nonspecific host defense mechanisms against enteric infection include pH, proteolytic enzymes and the detergent action of bile salts. Peristaltic movement of the gastrointestinal tract means an infectious agent will be swept away unless it has some adherence mechanism.

With *V. cholerae*, the mechanism of attachment to the mucosal surface is not known, but motility, together with the production of NA and proteases, is important. Once attached, a toxin is produced that binds to and enters the epithelium of the small intestine where it causes irreversible activation of adenylate cyclase raising cAMP levels. This results in inhibition of $Na^+$, $Cl^-$ and water uptake by villus cells and stimulated secretion of inorganic anions and water by crypt cells. Uncontrollable diarrhea results with huge loss of water and electrolytes.

Different strains of *E. coli* can be enterotoxigenic (watery diarrhea), enteroinvasive (dysentery, diarrhea), enterohemorrhagic (bloody diarrhea) and enteropathogenic (acute and chronic infant diarrhea). Pili bind to cell receptors which can be host specific and, in some cases, inhibited by mannose; mannose-resistant pili bind to neutral glycolipids. Enteroinvasive strains multiply within enterocytes, and adhesion of enteropathic strains leads to the loss of microvilli. Enterotoxigenic strains cause secretion of fluids and electrolytes into the lumen of the bowel whereas the other strains produce *Shigella*-like toxins.

In bacterial dysentery caused by *Shigella* spp. the outer membrane proteins of virulent strains are responsible for attachment to, and phagocytosis by, epithelial cells. During this intracellular phase the bacterium multiplies in the cytoplasm resulting in cell death. Shiga toxin possibly causes vascular damage with consequent inflammatory responses which result in diarrhea with feces containing blood, mucus and inflammatory cells. After attachment via pili, *S. typhi* also enters the cytoplasm of intestinal epithelial cells where ingested organisms are able to multiply. The toxic manifestations of typhoid fever are due to systemic infection with the release of LPS.

Rotaviruses are the most common cause of gastroenteritis in infants and are responsible for many deaths in developing countries. These viruses attach via their outer capsid protein and replicate in the differentiated columnar epithelial cells found at the apex of villi in the small intestine. Caliciviruses, responsible for gastroenteritis in both children and adults, cause the tips of villi to slough off. Coronaviruses may be responsible for gastroenteritis but more evidence is required.

The pathogenic properties of other enteric viruses also depend on attachment, and a cellular receptor of the immunoglobulin superfamily is used by poliovirus. This virus replicates primarily in the intestinal mucosa; less than 1% of poliovirus infections result in classic paralytic disease which can occur only after systemic spread followed by infection of the CNS. Hepatitis A virus also infects the intestinal epithelium before systemic spread.

The protozoan *G. lamblia* has a mechanical sucker to attach itself to epithelial cells in the duodenum and jejunum. This intestinal pathogen causes malabsorption which may be related to competition between the parasite and its host or to an effect of the parasite on villus cells.

### 3.1.4 Eye/conjunctiva

*Chlamydia trachomatis* (localized infections)
*Onchocerca volvulus.*

Of the human sense organs, the eye is most prone to infection. The conjunctiva are normally moist due to secretions from the lachrymal and other glands. Lysozyme, an enzyme with the ability to lyse bacterial cell walls, is present in tears but the mechanical wiping action of the eyelids is the main defense against infection.

Infection is essentially mechanical and may be effected by fingers or flies when the deposited infectious agents enter minor lesions on the conjunctiva. Specific attachment mechanisms may be utilized by *C. trachomatis* to adhere to the surface of conjunctival cells where intracellular growth causes cell death. This results in conjunctivitis and necrosis of hair follicles, with the resultant scars causing distortion of eyelids while the eyelashes wear away the corneal surface. Inflammatory responses take place so that the cornea becomes opaque. Interruption of the flow of tears enables secondary bacterial infections to occur leading eventually to blindness.

Onchocerciasis (river blindness) arises from within the infected host via systemic infection with microfilariae and localized inflammatory responses.

### *3.1.5 Genito-urinary tract*

*Haemophilus ducreyi* (localized infections), *N. gonorrhea, Treponema pallidum*
*C. trachomatis.*
HSV type 2 (localized infections)

The flushing action of urine will tend to prevent infection, and urine is usually sterile. In terms of their size, microorganisms also have a long journey to reach the bladder. The urethra is about 20 cm long in men but only some 5 cm in women, which explains why urinary tract infections are more frequent in women. Protection of the vagina against infection is achieved by the acid pH which results from the metabolic activity of a resident commensal lactobacillus in women during their reproductive life.

*T. pallidum* attaches by its tapered end and binds to fibronectin, the adhesive protein found on the mucosal surfaces of epithelial cells. This organism may possibly penetrate the skin and, in males, it is present in lesions on the penis but also in deeper genito-urinary sites. In females, *T. pallidum* causes lesions in the perineal region, labia, vaginal wall and cervix. *H. ducreyi* also causes ulcerative lesions (chancroid) in the paragenital areas. *N. gonorrhea* uses pili to attach to epithelial surfaces during the early stages of infection but pilin genes are 'switched off' later to prevent disadvantageous attachment to phagocytic cells.

In addition to eye infection by 'ocular strains', there are other strains or biovars of *C. trachomatis* causing genital infections. These include nongonococcal urethritis in males and mucopurulent endocervicitis in females. These infections may progress to more severe disease, lymphogranuloma venereum with abscesses in lymph nodes or, in females, infection of the upper genital tract with tubal scarring and pelvic inflammatory disease.

HSV type 2 is also passed by sexual intercourse, and glycoproteins in the viral envelope attach to receptors on epithelial cells. Herpetic lesions occur on the vulva, vagina, cervix, urethra and perineum of females and on the penis of males; recurrent infections are due to reactivation of latent infections.

## 3.2 Localized infections

Following their entry into the body, certain pathogens remain at the epithelial surface. This is sufficient for their multiplication and subsequent transmission to new hosts. Damage to the host is restricted to these sites unless soluble toxins are released to affect distant target organs or tissues. Some pathogens are restrained at epithelial surfaces by nonspecific host factors. For example, certain respiratory viruses replicate at 33°C in the upper respiratory tract but not at the internal body temperature of 37°C.

## 3.3 Systemic infections

A pathogen at its primary site of infection will be drained to the nearest lymph node where further growth or replication may occur. If a pathogen is not restricted to lymph nodes, it will eventually reach the bloodstream to be transported around the body in different compartments: free in plasma (poliovirus, yellow fever virus, JEV, dengue virus, HBV, pneumococci, trypanosomes); in erythrocytes (*Plasmodium*); in mononuclear cells (HIV, measles virus, herpesviruses including human cytomegalovirus (HCMV), *M. leprae, Leishmania*); in polymorphonuclear cells (staphylococci, streptococci); and in lymphocytes (EBV and HIV).

Direct introduction of a pathogen into the bloodstream can occur by various mechanisms: yellow fever virus, JEV, *Plasmodium* spp., *Leishmania* spp., *Trypanosoma* spp. and *Filariina* are injected by insect bites; rabies virus by animal bites; and HIV and hepatitis B and C viruses by transfusion of contaminated blood or intravenous drug abuse.

## 3.4 Organs or tissues targeted by pathogens following their systemic spread

### 3.4.1 Blood cells

Both red and white cells may be involved: erythrocytes (*Plasmodium*); lymphocytes (EBV and HIV), monocytes (mumps, measles, *L. donovani*) and resident macrophages in spleen, liver, bone marrow (*L. donovani*, dengue virus).

### 3.4.2 Liver

Yellow fever virus, hepatitis viruses, *Plasmodium* and *S. typhi* typically cause hepatic disease. *S. typhi* enters intestinal lymphatics and is also transported in the bloodstream to become localized in Peyer's patches, kidneys, gallbladder and, occasionally, bone marrow. Ulceration of Peyer's patches may give rise to intestinal perforation or hemorrhage which can be fatal. Yellow fever also has properties of viral hemorrhagic fever causing the massive gastrointestinal hemorrhage that results in 'black vomit'. Liver enlargement occurs in visceral leishmaniasis with hyperplasia of Kupffer cells which contain proliferating amastigotes. Destruction of hepatocytes is extensive in infections by hepatitis viruses and *Plasmodium*.

### 3.4.3 Spleen

*T. cruzi* amastigotes proliferate in splenic macrophages and visceral leishmaniasis also involves spleen. Anemia can occur due to hemolysis of erythrocytes by *Plasmodium* and by infiltration of bone marrow by macrophages containing *Leishmania* in late visceral leishmaniasis which displaces normal marrow components.

### 3.4.4 Central nervous system

Many different pathogens target the CNS, including JEV, rabies virus, *Plasmodium, Trypanosoma, Schistosoma,* measles virus (encephalomyelitis, subacute sclerosing panencephalitis (SSPE)), mumps virus (aseptic meningitis), *H. influenzae* and *N. meningitidis* (meningitis). Aseptic meningitis may be caused by viruses (poliovirus, mumps virus) entering cerebrospinal fluid from the blood and crossing into the brain through the ependymal lining. This route may also be used by other pathogens including *N. meningitis, M. tuberculosis, H. influenzae* and *S. pneumoniae.* Cerebral malaria results in acute meningoencephalitis and chronic encephalopathy. Schistosomiasis can affect the CNS due to focal masses of eggs surrounded by granulation tissue or to local growth of schistosomulae.

### 3.4.5 Muscle

Myocarditis is a common condition in *Trypanosoma* infections

### 3.4.6 Salivary gland

EBV, HCMV and mumps virus are found in saliva, which may be an important route of transmission of such infections.

### 3.4.7 Genito-urinary tract

In urinary schistosomiasis the deposition of eggs in the bladder, urine retention and formation of granulomas leads to secondary bacterial infections. Male genital filariasis can result in hydrocele, the accumulation of fluid around testes resulting in a grossly enlarged scrotum; chronic lesions of the testes and spermatic cord also occur. Genital filariasis in both males and females can give rise to chyluria (creamy white urine) due to lymphatic block and retrograde flow into the urinary tract. HSV type 2 is transmitted by sexual intercourse, and HIV present in the semen of infected males may infect the vagina during heterosexual intercourse or the rectum during male homosexual intercourse.

### 3.4.8 Placenta/fetus

Some systemic infections during pregnancy lead to transplacental infections that can have a devastating effect on the fetus which may be aborted or stillborn. In the first trimester, the fetus is especially susceptible to rubella virus infection which can cause cataracts, heart disease, deafness and mental retardation, but such risks decline rapidly in later pregnancy. The consequences of congenital HCMV infection are not so clearly related to the stage of pregnancy, and the consequences range from fatal to subclinical. An infected newborn may excrete large amounts of virus without any clinical signs of disease. If clinical signs do appear they reflect involvement of liver, kidney, lungs and CNS. HCMV infection may also be acquired during or immediately after birth from the infected mother excreting virus into the genital tract or in breast milk. Transplacental spread of *T. pallidum* results in congenital syphilis.

## 3.5 Oncogenic virus infections

Infection of B cells by EBV can have various consequences. Infectious mononucleosis or glandular fever is a disease that occurs mainly in young adults, and is typified by fever, sore throat, lymphadenopathy, splenomegaly, anorexia and lethargy. It occurs mainly in developed countries.

EBV infection may also be oncogenic, that is cause cancer. This virus is closely associated with Burkitt's lymphoma in African children and other B-cell lymphomas in immunodeficient patients. EBV is also linked very closely with NPC, a very common cancer in China and southeast Asia. Such geographical distribution suggests environmental factors are also important. Malaria is a cofactor for Burkitt's lymphoma, and components of the local diet, possibly nitrosamines in smoked fish, together with genetic factors affect susceptibility to NPC.

Other human viruses associated with cancer are hepatitis B virus, human T-cell lymphotropic virus type 1 (HTLV-1) and human papilloma virus.

## 3.6 Opportunistic infections

In addition to disease caused by 'aggressive' pathogens, there are also opportunistic infections caused by microorganisms that become pathogens only after a breakdown in host defenses. A classic example is puerperal fever (postpartum endomyometritis and septicemia), a common cause of death in the absence of suitable obstetric care following childbirth. This disease is caused by the invasion of soft tissues by bacteria that normally live on human skin and mucous membranes. The most common cause is *S. pyogenes*. *S. aureus* is particularly troublesome following injury to the skin by burns. *H. influenzae* in the lungs of influenza patients is a secondary bacterial infection that has taken advantage of the damage to tissue caused by a primary viral pathogen. A fatal disease concomitant with AIDS, *Pneumocystis carinii* pneumonia, is also caused by an opportunistic infection.

## 3.7 Transmission of infectious diseases

The different portals of entry into the human body may also be used by pathogens to exit from a diseased individual. Respiratory infections are spread by coughing and sneezing which eject droplets containing infectious agents that can travel at least 4 m through the air before they fall to the ground. Enteric infections are spread by the fecal–oral route; this may result from poor personal hygiene or from water supplies contaminated with human sewage. These mechanisms of transmission clearly explain why respiratory and gastrointestinal diseases are so prevalent. A related factor is the release of the pathogens responsible into the lumen of infected organs such as bronchioles in the lung or the ileum of the intestinal tract. Such release means these pathogens are kept away from most of the host's immune defenses apart from mucosal immunity.

Horizontal transmission of an infectious disease occurs directly from person to person by the routes described above but also by direct contact, particularly sexual intercourse. However, infection may also be acquired from infected animals or animal products (zoonoses); contaminated inanimate objects (fomites); or even as a result of medical intervention (iatrogenic and nosocomial infections). In some cases, an infectious or parasitic agent may also be passed from one generation to the next; such vertical transmission of infectious diseases is either congenital or acquired perinatally.

# Chapter 4

# Innate and acquired immune responses

Previous chapters have described the etiological nature of infectious and parasitic organisms, and the various means by which they can cause disease. Now it is important to consider the host's defense mechanisms in order to understand how they try to combat such infections.

The immune system of the human body can mobilize two main lines of defense. Innate or natural immunity is a rapid but non-specific response which includes inflammation, phagocytosis and activation of the complement system. Acquired or induced immune responses occur later, taking about 3–5 days to develop, and are effected through the specific recognition of foreign antigens. Antibodies (humoral response) identify structural components of extracellular pathogens or their toxic products, whereas specialized lymphocytes (cell-mediated response) can recognize host cells infected by intracellular parasites. To tackle the largest infectious agents, parasitic worms, both innate and acquired immune responses must act together.

Although they can be distinguished temporally, innate and acquired immune responses cannot be separated readily on a functional basis. For example, activation of the complement cascade can directly lyse pathogens but also yield factors that enhance antibody-mediated immunity. Phagocytic cells can both ingest pathogens and present foreign antigens to initiate cell-mediated immunity. Other molecular and cellular mechanisms facilitate and amplify these immune processes.

## 4.1 Innate responses to infection

### 4.1.1 Inflammation

Innate protection develops as soon as an infectious agent invades the host and injures host tissues (*Figure 4.1*). This results in the rapid release or production of several mediators of inflammation. These include the amino

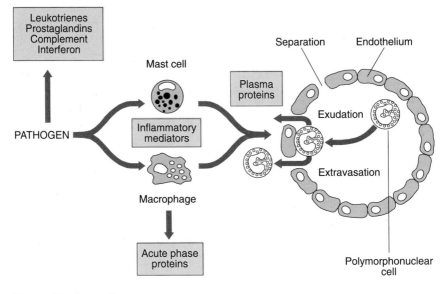

**Figure 4.1:** Innate immune responses.

acid, histamine; prostaglandins and leukotrienes, both products of arachidonic acid metabolism; polypeptides (kinins) produced by the action of enzymes called kallikreins on kininogen, a plasma $\alpha_2$-globulin; and components of the activated complement pathways. Some mediators are released from mast cells which are resident in body tissues.

Inflammatory factors act locally on blood vessels causing their dilation and increased permeability due to the formation of gaps between endothelial cells. This enables the extravasation or emigration of polymorphonuclear leukocytes (neutrophils) and monocytes (*Figure 4.2*). In the extravascular space, these leukocytes have directional movement or chemotaxis in response to soluble chemicals released at the site of infection. Such chemoattractant activity, which may include products of

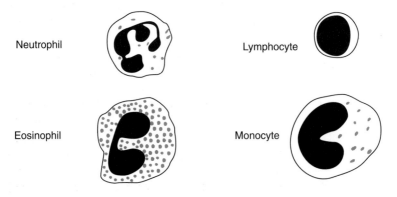

**Figure 4.2:** Schematic representation of leukocyte structure. Reproduced from Wardlaw, A.J. (1994) *Asthma*, BIOS Scientific Publishers, Oxford.

the infectious agent's own growth together with other, host factors, results in their arrival at the site of infection where monocytes mature into macrophages. Neutrophils are short-lived (3–4 days) due to apoptosis (programed cell death) but macrophages live longer, and in the later stages of an inflammatory response they become the principal phagocytic cell. Resident mononuclear phagocytes (alveolar macrophages) guard the lung against localized respiratory infection while specialized phagocytic cells guard the liver (Kupffer cells), brain (microglial cells), bone (osteoclasts) and kidney (mesangial macrophages), and other macrophages in the lymph nodes and spleen protect against sytemic infections.

The eosinophil, another leukocyte found in inflammatory exudates, is so called because it has cytoplasmic granules which bind acid dyes such as eosin (*Figure 4.2*). Eosinophils are weakly phagocytic cells but their cytoplasmic granules contain proteins that are particularly toxic for helminths when released extracellularly.

### 4.1.2 Phagocytosis

Phagocytosis of bacteria by macrophages was first demonstrated in 1882 by Metchnikoff. In 1904, Neufeld and Rampeau showed the phagocytic uptake of pneumococci was enhanced if antibodies were attached to the bacterial surface, a process called 'opsonization'. This is now known to be due to the presence on the surface of neutrophils and macrophages of specific receptor molecules (see Sections 4.1.3 and 4.3.1). Such observations led to the recognition of intracellular killing of bacteria and other microorganisms by phagocytic cells.

Ingestion of microorganisms into the cytoplasm of neutrophils or macrophages results in the formation of a phagosome which then fuses with a cytoplasmic organelle, the lysosome. Lysosomes contain numerous degradative enzymes including phosphatases, nucleases, lysozyme, collagenase and elastase. Development of the phagolysosome stimulates the hexose monophosphate shunt leading to the 'respiratory burst' which provides electrons necessary to produce the superoxide anion $O_2^-$, hydroxyl radical $OH^-$ and hydrogen peroxide $H_2O_2$. Superoxide dismutase converts the superoxide anion into hydrogen peroxide, a powerful bactericide. Another enzyme, myeloperoxidase, a lysosomal enzyme found in neutrophils but not in macrophages, acts with halides ($Cl^-$ and $I^-$) to form hypohalite. This bactericidal product damages bacterial cell walls by halogenation and by oxidation of amino acids to produce aldehydes. Lactoferrin, an iron-binding protein with antimicrobial activity, and other cytotoxic cationic proteins that bind to bacterial cell walls are also found in neutrophils but not macrophages. Macrophages have another mechanism of intracellular killing mediated by nitric oxide, a toxic chemical generated by the action of nitric oxide synthase on arginine.

The release of degraded phagolysosomal contents from phagocytic cells enhances chemotaxis and increases the inflammatory response. This is further amplified by activated macrophages and, during the adaptive immune response, by antigenically stimulated lymphocytes through their release of chemical mediators that provide an important link between innate and induced immunity (see Section 4.7).

### 4.1.3 Complement

Complement is the term coined by Paul Ehrlich to describe the activity in serum which, in the presence of specific antibodies, lyses bacteria. This is the 'classical pathway' of complement activation initiated by immune complex formation between antibodies and antigens. More recently an 'alternative pathway' has been recognized which can be initiated in the absence of induced immune responses. This is triggered directly by certain molecules, polysaccharides or lipopolysaccharides found on the surfaces of certain bacteria, enveloped viruses and virus-infected cells. Both complement pathways are a cascade of activities, each step being triggered by a preceding event (*Figure 4.3*).

The classical pathway begins when antibodies combine with antigens to form immune complexes which can bind avidly with a subunit of the first complement component, C1q. A conformational change effected by such complex formation activates another subunit, C1r, and, in turn, C1s, to produce an active serine protease. This enzyme acts on the next component, C4, to yield C4a and C4b. The latter is easily hydrolyzed and very unstable but, if C4b is bound to a cellular membrane, it acts as a binding site for another complement component, C2. Surface-bound C4b can then act on C2 to produce C2a and C2b. The resultant complex C4bC2a is known as the classical pathway C3 convertase enzyme, another serine protease, which cleaves C3 into C3a and C3b.

Activation of the classical complement pathway may also occur in the absence of antibodies. During the course of many diseases involving acute inflammation there is a rapid increase in the concentrations of certain plasma proteins as part of the so-called 'acute-phase response'. This includes the C-reactive protein, a β-globulin synthesized by hepatocytes, which forms complexes with surface polysaccharides found on a number of bacteria and parasites. This triggers the classical complement pathway with an efficiency equal to its activation by antigen–antibody complexes.

The alternative complement pathway is based on the slow hydrolysis of native C3 in plasma. The product, C3i, acts as a binding site for Factor B (FB); FB bound to C3i is cleaved by Factor D (FD), another serine protease, to Ba and Bb. The fluid phase C3iBb is another C3 convertase which cleaves C3 to C3a and C3b. Surface-bound C3b, which may also be produced by the classical pathway, can act as an initiation site for more FB, and FB bound to C3b is a further substrate for FD. Bb combines with

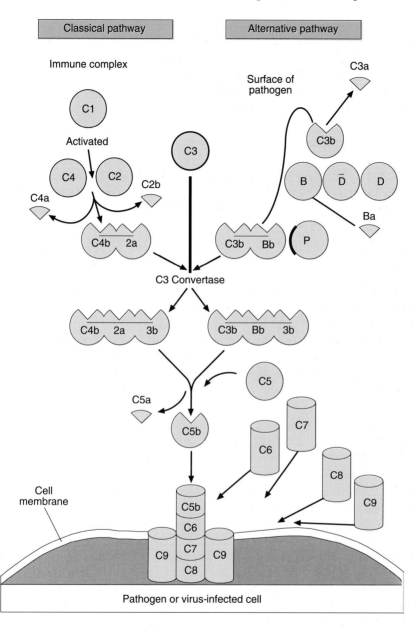

**Figure 4.3:** Classical and alternative complement pathways. Reproduced from Schaechter *et al.* (1993) *Mechanisms of Microbial Disease* with permission from Williams & Wilkins. © Williams & Wilkins, 1993.

C3b and the C3bBb enzyme cleaves more C3 but this complex dissociates rapidly in free solution. However, it is stabilized by various surface factors, including bacterial polysaccharides and viral envelope glycoproteins, which act, therefore, to initiate the alternative complement pathway. Another plasma protein, properdin (P), interacts with surface-bound

C3bBb to form the stable convertase, C3bBbP. This acts on C3 and C5 in a manner analogous to the classical complement pathway C3 convertase.

Complement fragment, C3b, facilitates opsonization via complement receptors, CR1 and CR3, found on the surfaces of phagocytic cells. Thus, activation of the alternative complement pathway and opsonization via C3b receptors is an important function of the innate immune response. Enhanced phagocytosis as part of the adaptive immune response can occur both directly via Fc receptors and through C3b receptors following activation of the classical complement pathway by immune complexes. The generation of C3a and C5a at the site of infection amplifies these mechanisms because both fragments are chemotactic for neutrophils and monocytes and also activate mast cells.

The final phase of the complement cascade – formation of the membrane attack complex (MAC) – is common to both classical and alternative pathways. It is initiated by C3b which combines with the C3 convertases appropriate to each pathway and cleaves C5 to C5a and C5b. The association of C5b with C6 and C7 forms an intermediate complex with a membrane-binding site; this complex binds C8 and part of the bound C8 molecule penetrates the membrane. Finally, C9 molecules aggregate into the complex with the formation of conical holes or pores in the membrane. Small molecules, water and salts, enter the cell with resultant damage to membranes and/or osmotic shock. This can disrupt enveloped structures such as virus particles, virus-infected cells, bacteria and parasites (*Figure 4.3*).

In order to prevent undesirable action against host cells, complement activation is tightly controlled. Mammalian cell surfaces are protected by the binding of C3b via complement receptors and/or membrane cofactor protein which favor binding of a regulatory inhibitor of the alternative pathway. The surfaces of bacteria, virus-infected cells and other parasites lack these features, and C3 deposition is unaffected. Host tissues are also protected from the lytic effects of MAC formation by the homologous restriction factor (HRF). This blocks C8 binding with C9, and vitronectin, a serum protein related to fibronectin and laminin, binds the C5bC6C7 complex to prevent its insertion into host membranes. Pathogens can, therefore, become specifically or selectively targeted for opsonization and for cytolytic attack.

## 4.2 Acquired immune responses to infection

There are several unique features of acquired immune responses to infection. First, they are specific: antibodies recognize the shape or conformation of target molecules, whereas the cellular response by specialized lymphocytes is determined more by their primary structure. Second, acquired immunity is selective, thus enabling discrimination between the pathogen and its host. Finally, there is memory. This ability to recall earlier encounters with infectious and parasitic agents facilitates long-term protection. A substance capable of inducing acquired immune responses is an antigen or immunogen.

## 4.3 Humoral immunity

### 4.3.1 Immunoglobulins

An antibody is a serum protein molecule able to bind specifically to antigen. Immunoglobulin (Ig) is the term used now to identify these immunologically reactive proteins that are found in the γ-globulin fraction of plasma. Ig can be divided into five classes or isotypes: IgA, IgD, IgE, IgG and IgM (*Figure 4.4*).

IgG is a Y-shaped molecule composed of two identical heavy chain (mol. wt 50–70 kDa) and two identical light chain (mol. wt 22 kDa) molecules. Digestion of IgG by the proteolytic enzyme, papain, cleaves at the neck of the IgG molecule to produce two identical fragments, one capable of recognizing antigen (Fab) and another that is readily crystallizable (Fc). In all Ig classes the amino acid sequences in the Fab fragment are variable (the V region). Within the V regions are

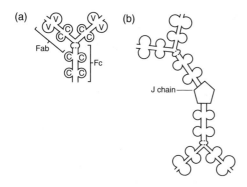

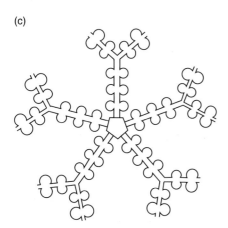

**Figure 4.4:** Structures of immunoglobulin molecules. (a) IgG, (b) IgA (dimer), (c) IgM.

hypervariable regions that bind with antigens and are responsible, therefore, for immunological specificity. Sequences in the remainder of the IgG molecule are constant (the C region). The Fc portion determines the biological properties of the particular class of Ig.

IgG molecules are monomeric (mol. wt 150 kDa). The Fc portion binds to Fc receptors found on the surfaces of macrophages, neutrophils and other cells. Activated complement component, C3, also binds to the Fc portion of IgG, and phagocytic cells also have C3 receptors. Thus, these mechanisms facilitate antibody-mediated opsonization. IgG associated specifically with surface antigens can also activate the lytic capability of the complement cascade. IgG can be divided into four subclasses which differ in their ability to fix complement: IgG1 and IgG3 are more effective than IgG2, but IgG4 is ineffective.

The IgM molecule is pentameric (mol. wt 970 kDa) with each individual molecule linked together by a joining or J chain. IgM binds particularly to any structure composed of identical repeating subunits, such as the surface of a virus particle or a bacterium, and is a very efficient activator of complement. IgM is produced following the first exposure to an antigen and is typical of this primary response. Further exposure to the same antigen results in the secondary response which progressively replaces IgM by other isotypes, a phenomenon known as isotype or class switching (see Sections 4.5 and 4.7).

IgA monomer (mol. wt 150 kDa) is most important as a secretory antibody. IgA1 is found predominantly in serum but the IgA2 subclass is found particularly at mucosal surfaces. The monomer has a structure similar to IgG but functionally IgA consists of a dimer joined by the J chain. A secretory component, derived from epithelial cells, is attached to facilitate transport and to protect against proteolytic enzymes in the lumen of the gut and at other mucosal surfaces. IgA2 is more resistant than IgA1 to proteases produced by pathogenic bacteria.

IgD (mol. wt 180 kDa) accounts for $\leqslant 1\%$ of total serum Ig but is found in large quantities on the surface of B cells. IgD does not have a known effector function.

IgE (mol. wt 190 kDa), found only in small amounts as serum Ig, is involved in anaphylaxis or immediate hypersensitivity (hay fever, asthma). In the presence of antigens (allergens) interaction of IgE with high-affinity Fc receptors on basophils and mast cells results in the release of histamine, prostaglandins and leukotrienes together with other cytokines and colony-stimulating factors. Eosinophils have low-affinity Fc receptors and these cells, together with IgE, are particularly important in protection against helminth infections.

### 4.3.2 B lymphocytes

Ig is secreted by plasma cells. Their precursors are B lymphocytes which originate from pluripotential 'stem cells' in the fetal liver or adult bone marrow. Mature B cells have receptors on their surfaces that are antibody molecules; all carry the monomeric form of IgM; some 70% also carry IgD, although the mature plasma cell lacks all surface immunoglobulin. Each B lymphocyte carries a specific Ig receptor for one particular antigen (*Figure 4.5*). Binding with an antigen activates this specific cell which undergoes rapid proliferation (clonal expansion). This results in the production of antibodies, a process which is facilitated by other lymphocytes known as T helper cells (see Sections 4.5 and 4.6).

Specific Ig is produced mainly in the spleen and lymph nodes and in mucosa-associated lymphoid tissue (MALT), an important site for IgA production. The IgM class of antibody is made initially, but continual antigenic stimulation can result in isotype switching (see also Section 4.7). The class of Ig produced is determined also by the site of antigenic stimulus – IgA, IgG and IgM are found in plasma but IgA is found particularly at epithelial surfaces – and by the type of antigen – IgE is especially induced by helminths.

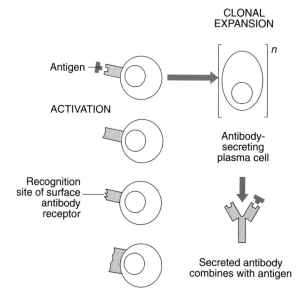

**Figure 4.5:** Specific activation by antigen and clonal expansion of B lymphocytes.

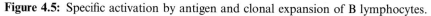

## 4.4 Cell-mediated immunity

### 4.4.1 The lymphoid system

The lymphoid system consists of lymphocytes (B and T cells) and accessory cells (phagocytic cells and antigen-presenting cells). Primary lymphoid tissue is found in discrete organs: B cells mature in the bone marrow and fetal liver; T cells mature in the thymus. Secondary lymphoid tissues make up the sites where cell-mediated and humoral immune responses are generated. They are again found in specialized organs – the spleen which is the source of B and T lymphocytes destined to enter the blood circulation, and lymph nodes, part of a network of lymphatic vessels which collects fluid draining from interstitial tissues. MALT protects against pathogens entering the body at epithelial surfaces. Such tissues are found lining the respiratory tract (bronchial-associated lymphoid tissue, BALT), the intestinal tract (gut-associated lymphoid tissue, GALT) and around the genito-urinary tract. MALT is responsible primarily for the production of secretory IgA.

## 4.5 T lymphocytes

T lymphocytes originate in the bone marrow as stem cells but their differentiation takes place in the thymus. T cells do not produce antibodies but associate directly with antigens on the surface of other cells. They recognize antigens by the T-cell antigen receptor (TCR). TCR-2 is a heterodimer of two disulfide-linked polypeptide $\alpha$ and $\beta$ chains. TCR-1 is structurally similar but consists of $\gamma$ and $\delta$ polypeptide chains; 95% of T cells express TCR-2 and 5% TCR-1 (*Figure 4.6*). Both $\alpha$ and $\beta$ chains of the TCR-2 heterodimer have variable and constant regions that are comparable with Ig molecules. These similarities in amino acid sequence and conformation reflect their common function of binding to a wide diversity of foreign antigens.

T cells carry other molecules known as cluster designations (CD) that can be used as surface markers. CD2 and CD3 are found on all T cells but CD4 and CD8 are found only on some subpopulations.

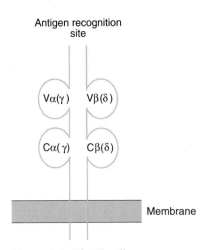

**Figure 4.6:** The T-cell receptor.

The major histocompatibility complex (MHC) molecules are other surface molecules that are also important in the interaction of T lymphocytes with their target cells (*Figure 4.7*). Class I MHC molecules are found on all nucleated cells but class II MHC molecules are expressed mainly on macrophages, B lymphocytes and activated T lymphocytes. MHC molecules are genetically determined and the MHC gene complex is located on human chromosome 6. These genes are very polymorphic, that is they vary at a high frequency in the human population. To date, more than 100 distinct class I and class II MHC molecules have been identified. Each variant of a polymorphic gene is known as an allele, and individuals sharing the same alleles are termed syngeneic whereas individuals with different alleles are described as allogeneic. The total set of MHC alleles is known as the MHC haplotype.

Class I MHC molecules consist of two polypeptide chains, an MHC-coded $\alpha$ chain and a $\beta$ chain that is encoded by a gene outside the MHC complex. The $\beta$ chain is identical with a protein initially found in urine and is called $\beta_2$-microglobulin. The native conformation of class I molecules depends on this association, and denaturation occurs if $\beta_2$-microglobulin is displaced. Class II MHC molecules consist of distinct $\alpha$ and $\beta$ polypeptide chains held together by noncovalent bonds. Both class I and class II MHC molecules are anchored firmly on cell surfaces and not secreted.

There is a deep cleft on the top surface of the extracellular portion of the MHC molecule (*Figure 4.7*). This cleft is lined by many of the amino acid residues that vary between alleles, and probably influences the specificity of peptide binding or the conformation of bound peptides. Within the cleft are noncovalently bound peptides which have been processed in different ways before they reach the cell surface. Proteins synthesized in the cytoplasm are degraded to peptides, and each peptide

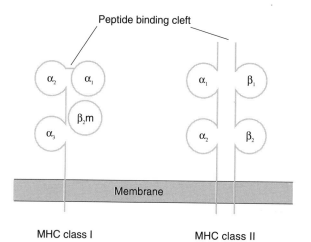

**Figure 4.7:** A schematic representation of MHC molecules.

associates with a class I MHC molecule. The $\beta_2$-microglobulin molecule is essential for formation of the stable complex to be expressed on the cell surface. Such class I MHC–peptide complexes are normally derived from cell protein, but they can contain viral peptides in virus-infected cells. Class II MHC molecules associate with peptides in endosomes; this association is most likely to occur if the protein has been taken into the cell by endocytosis. Such proteins may be the soluble products of a particular pathogen or even produced after phagocytic degradation of the pathogen itself.

T lymphocytes bearing TCR-2 can recognize peptides presented in the cleft of MHC molecules. CD4$^+$ T cells recognize peptides in association with MHC class II molecules whereas CD8$^+$ T cells recognize peptides in association with MHC class I molecules. During their maturation, these cells have been selected by clonal deletion of self-reactive T cells. This has several important consequences. First, T lymphocytes can recognize foreign antigens only on the surface of cells and, unlike antibodies, cannot recognize soluble antigens. This localizes the cytotoxic effects of T cells and prevents their inadvertent expression. Second, the foreign antigen is recognized only if the target cell has the same MHC haplotype as the T cell – a phenomenon known as MHC restriction. This necessity for simultaneous recognition of both foreign antigen and MHC molecule focuses further the effects of these cells. Finally, the genetic polymorphism of MHC molecules determines the conformation of the cleft and, consequently, both the specificity and affinity of peptide binding in T-cell recognition. This means individuals with different MHC haplotypes can vary considerably in their ability to generate cell-mediated immune reactions against the same antigen. The immune response genes that determine an individual's ability to respond by antibody production to particular antigens also map to class II MHC loci (see Section 4.5.3).

### 4.5.1 Human leukocyte antigen (HLA) system

The different MHC molecules expressed on human leukocytes were recognized initially by their antigenic nature expressed in graft rejection. Consequently, the genes in the MHC of an individual are recognized as the human leukocyte antigen or HLA system. Three gene products, HLA-A, HLA-B and HLA-C, are encoded by the MHC class I gene complex and HLA-DP, HLA-DQ and HLA-DR are the three gene products of MHC class II.

### 4.5.2 Cytotoxic T lymphocytes

CD8$^+$ lymphocytes are able to lyse any nucleated cell expressing a foreign antigen on its surface: hence, they are called cytotoxic T lymphocytes (CTL). Virtually all human cells have class I MHC molecules, and this gives CD8$^+$ CTL a broad range of cellular targets. The TCR site is closely associated with the CD8 and CD3 complex (*Figure 4.8*). It should be noted that foreign peptide antigens on the target cell surface must have both MHC-binding and TCR-binding residues. CD3, after the antigen has bound to the TCR, acts as a transducer to activate the T cell. Activated CTL attach to the surface of target cells, and cytoplasmic granules release 'perforins' which are molecularly similar to the C9 component of the complement MAC. Perforins bind to the target cell membrane, polymerize and form transmembrane channels which cause cell lysis. CTL also produce a cytotoxin that fragments DNA in the target cell nucleus.

### 4.5.3 T helper cells and antigen presentation

CD4$^+$ lymphocytes, which recognize antigens presented in association with class II MHC molecules, are called T 'helper' (Th) cells. They cooperate with other cells that also express class II MHC molecules known as 'antigen-presenting cells' (APC). These are mainly phagocytic cells of the monocyte/macrophage lineage and are particularly effective in the processing of particulate antigens and immune complexes taken up via their Fc and C3 receptors. Other APC are dendritic cells, nonphagocytic cells derived from the bone marrow but of a different lineage to macrophages. Langerhan's cells are dendritic cells in the skin where they process antigens for local immune responses. These cells also migrate via the lymphatic system (in which they are called 'veiled' cells) to the lymph nodes where they present antigens to T cells. Follicular dendritic cells are

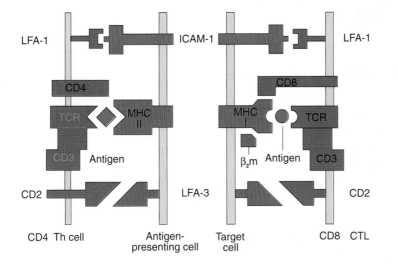

**Figure 4.8:** Molecular interactions between T lymphocytes, antigen-presenting cells and target cells for cytolytic activity.

found in germinal centres of lymph nodes with immune complexes bound to the dendritic processes to give a beaded appearance.

APC have two important functions: first, they convert protein antigens to peptides in association with class II MHC molecules which can then be recognized by Th cells. Antigens taken up by APC may be up to 1000 times more immunogenic than native antigen. Second, APC stimulate Th cell activity. The close association of APC and T lymphocytes is facilitated by integrins and adhesion molecules (*Figure 4.8*). Leukocyte functional antigen (LFA) molecules are integrins found on all leukocytes; LFA-3 binds with the T-cell membrane protein, CD2. Intercellular adhesion molecule (ICAM)-1, and ICAM-2, expressed on activated T cells bind with LFA-1. CD3 acts as a signal transducer activating second messenger systems leading to T-cell proliferation.

Th cells serve primarily to activate APC and to assist B lymphocytes in their production of antibodies. Th cells secrete a molecular signal (see Section 4.7) which causes induction of class II MHC molecules on APC resulting in more antigen presentation to T cells. Cooperation between Th and B cells is required for the production of antibodies against so-called 'T-dependent antigens' and also to effect class or isotype switching. The primary antibody response to an antigen by a B lymphocyte is IgM, but further antigenic stimulation in the presence of Th cell elicits the secondary response when IgG or IgA, but sometimes IgE, are produced (see Section 4.7). Consequently, class switching is shown only by T-dependent antigens, and T-independent antigens may elicit only IgM. T-independent antigens are typically polymeric with repeating antigenic determinants (such as LPS and bacterial flagellin).

A consequence of these different mechanisms of antigen recognition by $CD4^+$ and $CD8^+$ T lymphocytes is their selective activation, which provides a more effective defense against different pathogens. Extra-cellular pathogens may be phagocytosed by macrophages or produce soluble antigens that will be recognized by B lymphocytes. This results in class II MHC-associated antigen presentation and further cooperation with $CD4^+$ Th cells to produce an enhanced humoral immune response. Intracellular pathogens will give rise to class I MHC-associated antigens that are best recognized by CTL.

### 4.5.4 Natural killer cells

Lymphocytes other than CTL can kill target cells. Natural killer (NK) cells kill target cells without MHC restriction – and in the apparent absence of an induced immune response. The nature of the target receptor for NK cells is not known but possibly they recognize altered cell surfaces. NK cells have an important role in innate immunity, particularly during the first few days of virus infections. They are activated by chemical mediators released from other cells (see Section 4.7) to become lymphokine-activated killer (LAK) cells or large granular lymphocytes (LGL). NK cells can also combine via their Fc receptors with immune complexes on the surface of target cells in a process called antibody-dependent cell-mediated cytotoxicity (ADCC).

## 4.6 Antigens

Antigens elicit specific antibodies by recognition of Ig on the surface of B cells and induce $CD4^+$ or $CD8^+$ T-cell responses which occur through interaction of the TCR with foreign peptides held in MHC class II or MHC class I molecules, respectively.

An antibody binds to a particular part of an antigen – the antigenic determinant or epitope. An antigen may have several different epitopes but an epitope isolated from its carrier molecule may not elicit an immune response (a hapten). An epitope may consist of a linear sequence of amino acids (continuous epitope) or be formed only after one or more polypeptides have been brought together (conformational or discontinuous epitope). To form immune complexes the antigen and antibody molecules must come into close proximity, and their conformations must enable a tight fit. This reversible union is effected by hydrogen bonding, electrostatic forces, van der Waals bonds and hydrophobic bonds. The variable domain of the Fab fragment of the antibody molecule is responsible for specific antigen binding. X-ray crystallography of lysozyme–Fab complexes shows 17 amino acid residues of the antibody are apposed to 16 residues of the antigen. However, an antibody recognizes the overall shape of an antigen rather than its specific chemical composition. Such exquisite discrimination is shown experimentally by antibodies that can distinguish between the dextro- (D) and levo-rotatory (L) isomers of optically active molecules.

When mixed, antigen and antibody molecules combine to form complexes that constantly bind and then dissociate. The concentration of antigens which enables half of the antibodies to be complexed with antigens is a measure of the affinity or strength of molecular binding. Antibody affinity increases as the immune response progresses, being higher in a secondary than a primary response. This process of affinity maturation is due to somatic mutations of Ig genes and selective activation by antigens of B cells which produce antibodies with increased binding affinity with that antigen.

T-cell responses, as described above, are regulated by MHC molecules that bind and display small peptides derived from foreign antigens. The predominant length of peptides associated with class I molecules is nine amino acid residues, although peptides up to 14 residues have also been identified. The amino- and carboxyl-terminal ends of class I-associated peptides are held by conserved residues located in pockets, termed A and F, at either end of the binding site. The centrally located pockets B, C, D and E, together with other polymorphic residues in pocket F, play an important part in the process by which residues are bound. Most of the peptides recovered from class II molecules have 12–19 residues but the overall range is from 10 to 30 residues. Unlike class I molecules, the peptide-binding sites of class II molecules are open at each end with an overall peptide contact length of 15 residues in a fully extended conformation.

Epitopes recognized by B cells and by T cells are different. The native antigen molecule interacts with Ig on the surface of B cells but before it can induce a Th cell or CTL response the same molecule must be degraded to short peptide sequences. Thus, conformation or shape is important for antigen recognition by antibody but primary structure of the presented peptide is more important in T-cell recognition. The immunogenic portions of proteins appear to be regions with a mixture of hydrophilic and hydrophobic amino acids. These amphipathic regions may be either B-cell epitopes or T-cell epitopes. This can be demonstrated with synthetic peptides based on equivalent amino acid sequences that combine with specific monoclonal antibodies or stimulate T cells, respectively.

It is clearly important that an antigen should have both B-cell and T-cell epitopes. Antibody production by B cells is effected by activated Th cells that have simultaneously been specifically stimulated to produce chemical mediators that facilitate B-cell growth and differentiation (see Section 4.7).

## 4.7 Cytokines

Both innate and induced immunity are modulated by chemical mediators called cytokines (*Figure 4.9*). These protein hormones also effect

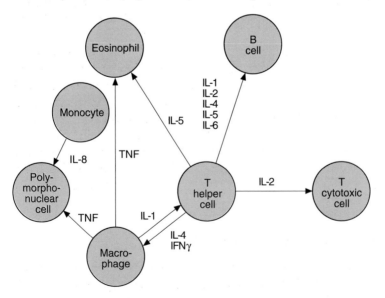

**Figure 4.9:** The cytokine network.

considerable interaction between the innate and adaptive immune responses. A cytokine produced by a leukocyte that acts upon a leukocyte is called an interleukin (IL).

### 4.7.1 Cytokines and innate immunity

Interferon (IFN) was first recognized as an antiviral factor produced by cells after virus infection. IFN-$\alpha$, obtained from leukocytes, or IFN-$\beta$, produced by fibroblasts, can protect neighboring uninfected cells against further virus infection by induction of cellular enzymes that inhibit viral transcription and translation. These IFNs can also increase the lytic capability of NK cells, so enhancing another innate mechanism of host defence.

Tumor necrosis factor $\alpha$ (TNF-$\alpha$) is produced mainly by macrophages stimulated by exposure to LPS molecules present on or released from the surface of Gram-negative bacteria. This cytokine was first recognized in serum taken from humans with bacterial infections by its necrotic activity against tumors, and as a mediator of a wasting syndrome (cachexia) consequent on chronic infectious disease. TNF acts on vascular endothelial cells to increase their adhesiveness to neutrophils and, subsequently, for monocytes and lymphocytes, to facilitate extravasation. TNF also activates phagocytic cells, including eosinophils, to enhance their intracellular killing. Activated macrophages, together with T cells and NK cells, are stimulated to produce more TNF and other cytokines including IL-1, IL-6 and IL-8.

Lymphotoxin (LT or TNF-$\beta$) shares many properties with TNF-$\alpha$ but it is produced only by activated T cells. However, TNF-$\beta$ acts locally whereas TNF-$\alpha$ is made in greater quantities and acts systemically, sometimes with toxic effects.

IL-1 is also produced mainly by activated macrophages during an innate response. IL-1 receptors are found on many different cell types and ligand binding enhances the action of IL-1 on macrophages and endothelial cells to produce IL-1, IL-6 and IL-8. IL-6 acts with TNF and IL-1 to augment the synthesis of acute phase proteins, such as C-reactive protein, by hepatocytes. IL-8 is produced by epithelial cells, fibroblasts, endothelial cells, monocytes and activated T cells in response to LPS, TNF-$\alpha$ and IL-1. IL-8, a 'chemokine', is a chemotactic factor for neutrophils and other lymphocyte cells and aids extravasation by facilitating leukocyte adhesion to endothelial cells.

### 4.7.2 Cytokines and adaptive immunity

In addition to their antiviral properties, IFN-$\alpha$ and IFN-$\beta$ increase expression of class I MHC molecules but downregulate class II MHC expression. The former facilitates cytotoxic T-cell activity but the latter hinders Th-cell function. Such modulations favor recognition of virus-specific antigens by CTL, resulting in death of the infected cell, rather than their recognition by antibodies which would enable complement-mediated lysis to release progeny virions. Thus, production of these IFNs encourages immune responses to act locally and prevent further virus dissemination.

IFN-$\gamma$ is produced by T lymphocytes after their antigenic activation. This cytokine has antiviral activity but, unlike IFN-$\alpha$ and IFN-$\beta$, it has other properties that effect regulation of the immune response. IFN-$\gamma$ increases expression of both class I and class II MHC molecules on a wide variety of cell types; enhances expression of Fc receptors; and upregulates IL-2 receptor expression on T cells. These various effects facilitate the cognitive or specific phase of the immune response by increasing both antibody and cell-mediated immune responses. IFN-$\gamma$ also promotes innate immunity by activation of macrophages and NK cells.

Activated macrophages mediate specific immune responses by upregulation of the IL-2 receptors and consequent increased IL-2 production by CD4$^+$ T cells. IL-2 is produced by antigen-stimulated T cells and was known originally as 'T-cell growth factor'. (Another recently described cytokine, IL-15, also acts as a growth factor for T cells but is produced by other types of cell.) Because IL-2 receptors are found on T cells, IL-2 functions as an autocrine on the same T cell and as a paracrine on neighboring T cells. IL-2 also stimulates NK cells and LAK cells and acts synergistically with IL-5.

IL-4 is produced by Th cells and interacts variously with different lymphocyte populations. It is a growth factor for B lymphocytes and facilitates Ig class switching. At high concentrations it is a switch factor for IgE production while suppressing IgM, IgG2 and IgG3. IL-4 induces increased expression of CD23, a low-affinity receptor for the Fc portion of IgE, on resting B cells. IL-5 also affects B-cell differentiation. It increases IgA secretion by B cells although, unlike IL-4, it does not induce class switching but the maturation of B cells already committed to IgA production. IL-6 is the main growth factor for activated B cells and the major inducer of terminal differentiation leading to Ig production. IL-1 also augments the growth and differentiation of B cells.

### 4.7.3 Cytokines and Th cell control of immune responses

It has recently been recognized that there are two Th cell subsets, Th1 and Th2, characterized by their different cytokine profiles (*Figure 4.10*). Th1 cells produce TNF and IFN-$\gamma$ together with IL-2, whereas Th2 cells produce IL-4, IL-5, IL-6 and IL-10. The cytokines produced by the Th1 subset activate macrophages and mediate the development of CTL and ADCC. Conversely, the cytokines produced by Th2 cells effect B-lymphocyte proliferation and differentiation, Ig class switching resulting in the production of IgE and stimulate mast cells and eosinophils. IL-10 and

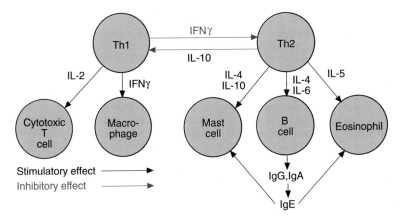

**Figure 4.10:** Modulatory effects of cytokines produced by T-helper cell subsets.

IFN-γ are mutually antagonistic: IL-10 produced by Th2 cells is especially effective in inhibiting the production of IFN-γ, and IFN-γ inhibits the growth of Th2 cells. Another recently described cytokine, IL-12, acts synergistically with IL-2 to abrogate the effect of IL-10.

Differential expression of Th1 or Th2 cell subsets appears to be determined by APC type and other, costimulatory signals. Immunization with certain antigens can elicit the Th1 subset if administered in solution, but the Th2 subset if in suspension.

## References

1. Schaechter, Medoff and Eisenstein (1993) *Mechanisms of Microbial Disease.* Williams & Wilkins, Baltimore.

Chapter 5

# Immune responses to infectious and parasitic diseases

Pathogenic organisms are clearly able to overwhelm both the innate and induced phases of host immunity because, for a time at least, they inflict such tissue and/or organ damage on their host that there are clinical signs of disease. At this stage, a crucial factor in further progress of disease is the pathogen's rate of replication, multiplication or reproduction measured against the speed of mobilization of the host's humoral and cellular immune responses. In an acute disease the pathogen gains only a brief advantage before it is eventually subdued by the development of effectual immunity (*Figure 5.1*).

With certain pathogens, however, initial infection may be followed by recurrent or chronic disease despite the induction of immune responses directed specifically against the pathogen (*Figure 5.1*). Clearly, such persistent infectious or parasitic diseases are caused by pathogens that are able to avoid or to evade the host's immune defenses, and this may be achieved by a variety of mechanisms. Some are able to modify their external features by either antigenic diversity or antigenic variation with the consequent requirement for different, specific immune responses following each new infection. Certain pathogens, after they have caused acute disease following primary infection, are not eliminated but remain as latent or dormant infections to inflict further damage on their host when reactivated at later times. Others deceive their host's immune responses through their antigenic complexity which presents a bewildering array of diversionary targets or by their ability to mask themselves with host-specific antigens.

This chapter describes the innate and induced immune responses that accompany infectious and parasitic diseases caused by the pathogens listed in *Tables 1.4* and *1.5*. Such knowledge may then be used to try to identify the location and nature of those immune responses that give

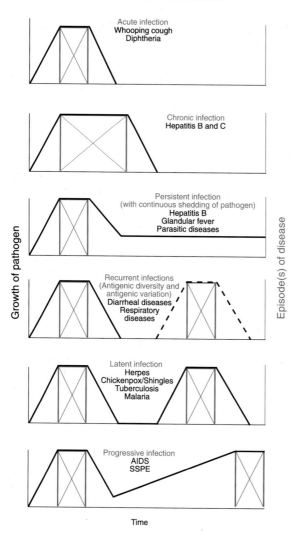

**Figure 5.1:** Disease patterns of different microbial or parasitic infections.

protection. It has been argued that vaccines should be designed to raise artificially those immune responses which protect against natural infection. Such a notion may be applicable to acute infectious or parasitic diseases but it is clearly inadequate in the case of recurrent or persistent diseases. More sinisterly, some pathogenic organisms may induce innate or acquired immune responses which exacerbate disease by promoting infection or causing damage to the host.

## 5.1 Innate immunity

The earliest responses of innate immune mechanisms may be triggered directly by external structural components of the pathogen itself. The alternative complement pathway is activated with the deposition of C5b enabling further complement components C6–C9 to set up the MAC. This lytic mechanism can directly destroy some enveloped viruses, virus-infected cells, certain Gram-negative bacteria and protozoa. Gram-positive bacteria may also activate this cascade pathway but they are protected by the long peptidoglycan molecules in their cell walls which hold the MAC away from the bacterial membrane. Other pathogenic bacteria avoid complement-mediated lysis by different mechanisms: *N. meningitidis* (group B) and *E. coli* (K1) capsular polysaccharides mimic host sialic acid and bind control factor H; the surface M protein of group A streptococci acts as an acceptor for factor H. The vertebrate stages of parasitic infections also avoid complement-mediated lysis by their acquisition of host regulatory factors, for example decay accelerating factor (DAF), or by shedding surface molecules.

Injury caused to host tissues by local infection and/or the release of bacterial exotoxins, particularly LPS by Gram-negative bacteria, results in the production of cytokines that trigger other innate responses. Chemoattractant activity mediated by IL-8 is augmented by prostaglandins and leukotrienes released from damaged host cells, and further supplemented by C3a and C5a generated by activation of the alternative complement pathway. Other cytokines, IL-1, IL-6 and TNF-$\alpha$ stimulate hepatocytes to synthesize and release acute-phase proteins: C-reactive protein triggers the alternative complement pathway after it has reacted with the capsules of certain bacteria (pneumococci) and other proteins neutralize bacterial toxins, including LPS. Virus-infected cells produce IFN-$\alpha$ or -$\beta$ which protects surrounding uninfected cells and restricts further spread of progeny virions. IFN release also enhances innate immune activity against infected cells by activation of NK cells. Some viruses, however, inhibit IFN production – such an inhibitor is found in cells infected with HIV, and only limited amounts of IFN are produced following HBV infection.

Inflammatory responses facilitate the extravasation of phagocytic cells and their migration into infected tissues. Phagocytosis is considerably enhanced by opsonization with C3b derived from the alternative complement pathway which may already be attached to the surfaces of pathogens or infected cells. However, certain pathogens are able to avoid phagocytosis or even mount a counter-offensive against phagocytic cells. Capsulated bacteria, classically the polysaccharide capsules of pneumo-cocci, resist phagocytosis, and similar resistance is also exhibited by smooth strains of *E. coli*. The M protein of *S. pyogenes* is another important anti-phagocytic factor. Other pathogenic streptococci produce streptolysin O and streptolysin S which lyse membranes, including lysosomes, with the release of lysosomal enzymes into the cytoplasm of phagocytic cells. *S. aureus* produces a leukocidin which effects the discharge of lysosomal contents to the exterior with resultant damage to surrounding tissue. The damage to cell membranes caused by streptolysin O, cholera and pertussis toxins inhibits polymorphonuclear chemotaxis at concentrations lower than their lytic effect.

If phagocytosis does occur, some engulfed pathogens survive or even multiply within phagocytic cells by the use of other countermeasures. Ingested catalase-positive staphylococci avoid intracellular killing by breaking down hydrogen peroxide produced by the myeloperoxidase pathway; *N. meningitidis* and *Leishmania* amastigotes express superoxide dismutase. Virulent *M. tuberculosis* prevents fusion of lysosomes with phagosomes and there is a correlation between its virulence and soluble factors extracted from these bacteria: 'cord factor' and sulfatides (trehalose-2′-sulfate esterified with long-chain fatty acids). *M. leprae* escapes from the phagosome into surrounding cytoplasm before lysosome–phagosome fusion can occur, and trypanosomes also get out of phagolysosomes to multiply in the cytoplasm of macrophages. Certain enveloped viruses (herpesviruses, measles virus) can replicate in phagocytic cells either because they enter directly into the cytoplasm by membrane fusion rather than inside a phagocytic vesicle or, if phagocytosed, they exit quickly from the phagosome before it fuses with a lysosome.

Activated macrophages are a source of cytokines – IL-1, IL-6 and TNF-$\alpha$ – that activate other macrophages and increase production of acute-phase proteins. IL-1 and IL-6 will also act later on T and B cells which, together with their role in antigen presentation, make phagocytic cells an important link between innate and acquired immune responses.

The ability to engulf extracellular particles is a primitive one, and it is not restricted solely to highly specialized cells. Phagocytosis by epithelial cells may actually provide a mechanism for transfer of pathogens to sub-mucosal spaces. Invasive *Neisseria* spp. are transported by nasopharyngeal epithelial cells; *Shigellae* and enteroinvasive *E. coli* by the intestinal mucosa; and *C. trachomatis* by ocular and genito-urinary epithelia.

Phagocytosis by neutrophils and macrophages is particularly important as a defense against bacterial infections. This is reflected by recurrent bacterial infections in two rare inherited diseases characterized by the inability to effect intracellular killing by these phagocytic cells. Chronic granulomatous disease arises from NADPH deficiency with consequent inability to produce superoxide ions. In Chédiak–Higashi disease, another autosomal recessive disorder, there is premature fusion of neutrophil granules during their development in the bone marrow. C3 deficiency also results in extreme susceptibility to infections with pyogenic bacteria.

Normal inflammatory responses following infection can lead to suppuration (formation of pus) due to the large number of live and dead pathogenic organisms at the local site of infection. Accumulation of dead neutrophils, followed by their autolysis and release of lysosomal enzymes, increases the extent of the lesion. Abscesses are formed by bacteria on the skin (streptococci, staphylococci) when the deposition of fibrin encloses the infected area and pus accumulates. Abscesses may also develop on infected internal organs, for example, on the brain, generated by bacteria found normally in the mouth or oropharynx (staphylococci, streptococci, *N. meningitidis*).

Activation of innate immune mechanisms can contribute to the pathology of infectious diseases. This is seen especially in infections by Gram-negative bacteria resulting from the release of endotoxins, particularly LPS, although similar mechanisms are induced by staphylococci, *B. pertussis* and *Plasmodium*. The common link is the production of IL-1 and TNF by activated macrophages. Apart from their effects on the hypothalamus leading to fever, these cytokines increase vascular permeability with consequent hypotension and shock. IFN production may be responsible for the clinical symptoms of some viral infections; injection of purified IFN into volunteers produced fever, headache and muscular pain.

## 5.2  Acquired immunity

Humoral immunity can act directly against pathogens by a variety of mechanisms. IgA is especially important in the protection of mucosal surfaces against infection, although some bacteria (*N. meningitidis, H. influenzae, S. pneumoniae*) produce proteases that are able to degrade this secretory immunoglobulin. Antibodies can immobilize certain bacteria by interaction with their flagella but, more significantly, they block the attachment of pathogens to receptors on susceptible cells. Specific antibodies 'neutralize' virus infectivity either by preventing adsorption of the virion to host cells or by effecting steric changes in the virus nucleocapsid that inhibit uncoating and the release of viral genomes within infected cells. Some pathogens, particularly bacteria, may remain as localized infections at mucosal surfaces and release toxins which are transported systemically to act against distant target organs. In such diseases (diphtheria, pertussis, tetanus), circulatory IgM or IgG antibodies are directed against the toxin and secretory IgA antibodies against the bacterium.

Attachment of IgA to viruses and chlamydia can block their adsorption to host epithelial cells while IgM and IgG protect against their systemic spread. Specific antibodies may also impede motility and prevent attachment of extracellular, parasitic protozoa to target tissues and facilitate their complement-mediated lysis or engulfment by macrophages. However, larger, extracellular parasites, particularly helminths, are insusceptible to these defensive mechanisms due to their size and tough external coat. Another distinctive response is necessary for their elimination involving IgE, eosinophils and mast cells. Eosinophil chemotactic factors are released by the worm, and worm antigens directly stimulate IgE production which causes mast cell degranulation and releases factors enhancing further eosinophil infiltration. ADCC mediated by IgE receptors and the release of eosinophil major basic protein lyse the parasite.

Antibody molecules act specifically to strengthen further attacks against targets that have been marked out nonspecifically by earlier inflammatory responses (*Figure 5.2*). The Fab portion of IgG can combine with the antigenic determinants of the infectious or parasitic agent, while the distal Fc portion is able to combine with Fc receptors on the surface of phagocytic cells. Such opsonization enhances the phagocytosis of pathogenic organisms. However, this mechanism is subverted by protein A of *S. aureus* which binds IgG by the Fc region; there are an estimated 80 000 binding sites on each bacterium.

Although antibodies are usually unable to enter into viable cells, they can complex with the foreign antigens that are incorporated into the surfaces of cells infected with intracellular parasites – viruses, chlamydia, certain bacteria and protozoa. This facilitates attachment of the complex through the Fc receptors of neutrophils and macrophages. Infected cells are also susceptible to further, specific attack by killer lymphocytes through ADCC. These various effector cells can then either engulf the infected target cell or cause its destruction by the release of toxic granules. Herpesviruses may evade such mechanisms by increasing the expression of Fc receptors on the surface of infected cells.

Immune complex formation either on a pathogen's surface or on the surface of cells infected by intracellular parasites activates the classical complement pathway. Consequent generation of C3b facilitates further opsonization using C3b receptors on macrophages. Complement activation also leads to assembly of the MAC which can lytically destroy

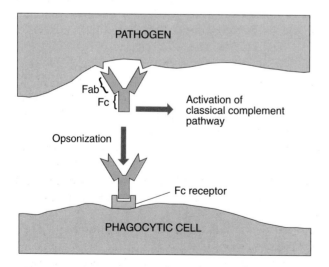

**Figure 5.2:** Antibody-mediated enhancement of innate immune responses.

extracellular enveloped viruses, chlamydia and Gram-negative bacteria. Although host cells infected with intracellular parasites are also susceptible to such lytic attack, the damage caused may be repaired by cellular mechanisms before lysis occurs.

Humoral responses to a pathogen are not always beneficial to the host. Opsonization may be counterproductive if antibodies are directed against pathogens that can multiply or replicate in phagocytic cells. The pathogen–antibody complex may be taken up by Fc receptors, thus facilitating the pathogen's entry into host cells or even enabling infection of cells that may not be susceptible to the pathogen alone. The former is illustrated by dengue virus infection where such 'immune enhancement' may occur. Primary infection induces antibodies that are cross-reacting but nonneutralizing with another dengue virus serotype. Consequently, the secondary infection is more severe due to enhanced uptake into phagocytic cells which are highly permissive for the replication of dengue viruses. Such antibody-enhanced infection is also seen in malaria, and HIV infection of mononuclear phagocytes may also be enhanced by opsonizing antibodies. Upregulation of Fc expression on the surface of herpes-infected cells might also enable these cells to become co-infected via immune-complexed HIV. This mechanism also enables HIV infection of cells that do not express CD4 receptors.

If large amounts of soluble antigens are produced during a pathogen's multiplication or replication, circulating immune complexes may be formed. Following their deposition in tissues, local inflammatory responses and activation of complement can occur with resultant local damage. Immune complex deposition in the skin after systemic infection can give rise to rashes (dengue) or to severe local necrosis (leprosy). Their deposition also leads to immunopathological disease in other major organs – the kidneys (malaria, trypanosomiasis, hepatitis B, dengue, leprosy), heart (bacterial endocarditis due to streptococcal infection), heart and brain (trypanosomes), liver (hepatitis B) and eyes (leprosy and measles).

The importance of specific humoral immunity in resistance to infectious and parasitic diseases is illustrated by the susceptibility to infection of individuals with congenital B-cell deficiencies leading to agammaglobulinemia or to selective immunoglobulin isotype deficiencies. The complete absence of γ-globulin from serum results in recurrent infections by pyogenic bacteria together with decreased resistance to certain virus infections, including poliovirus, and to intestinal parasites such as *Giardia*. Some patients with selective IgA deficiency have recurrent respiratory tract and enteric infections that lead to permanent damage, although other patients are apparently unaffected.

Cell-mediated immunity is the more important acquired response to infection by intracellular parasites. CTL recognize and destroy virus-infected cells if they express foreign antigens on their surfaces. Such immune responses prevent virus replication, whereas humoral immunity can merely act against extracellular dissemination of virus. However, recognition of infected cells by CTL can also lead to immunopathological damage, for example certain virus rashes and destruction of hepatocytes in HBV infection. With the larger, protozoan parasites – *Giardia, Trypanosoma, Leishmania* and *Plasmodium* – CTL are directed specifically against both extracellular and intracellular stages in their life cycles.

Antigenic stimulation of both $CD4^+$ and $CD8^+$ T cells promotes the release of IFN-$\gamma$ which activates macrophages to enhance their killing of intracellular parasites, particularly *M. tuberculosis, Trypanosoma, Leishmania* and *Plasmodium,* together with direct inhibition of the multiplication of malaria parasites in hepatocytes. This cytokine also enhances another element of innate immunity to target virus-infected cells by NK cells, possibly by their recognition of altered MHC molecules.

Chronic antigenic stimulation by persistent infection with intracellular parasites can lead to the formation of granulomas (*Figure 5.3*). These lesions occur during such diseases as tuberculosis or leprosy and in some parasitic infections. Granuloma or tubercle formation in lymph nodes during chronic tuberculosis is due to initial infection of macrophages by *M. tuberculosis.* Infected macrophages continue to grow, a process that is further encouraged by cytokines released during subsequent

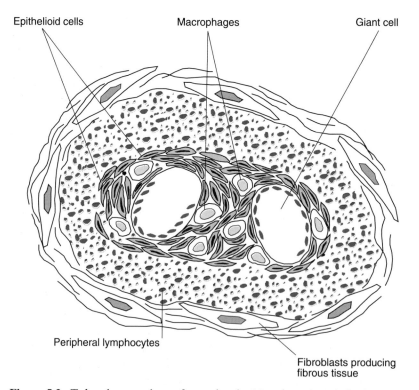

**Figure 5.3:** Tubercle-granuloma formation in *M. tuberculosis* infection. Reproduced from ref. [1] with permission from Blackwell Science Ltd.

cellular immune responses. The activated macrophages form elongated epithelioid cells which fuse to produce giant cells. Fibrous tissue forms around the tubercle and a necrotic center develops (caseation) which, if it liquifies, releases large numbers of bacteria that multiply vigorously resulting in the rapid spread of organisms to many organs including the liver, spleen, lungs and brain. Because *M. tuberculosis* is an obligate aerobe, its rapid growth in the lung can quickly give rise to fatal bronchopneumonia.

Mycobacteria also activate T cells with the $\gamma\delta$ TCR due to their expression of antigens that are homologous to 'heat-shock proteins'. These evolutionarily conserved molecules also activate $\gamma\delta$ T cells when either prokaryotic or eukaryotic cells are exposed to physiological stress. Certain bacterial toxins, for example staphylococcal enterotoxins, stimulate large numbers of $CD4^+$ T cells such that all the T cells in an individual that express particular $V_\beta$ TCR genes are activated. Such production of bacterial 'super-antigens' will also result in excessive cytokine production. Both mechanisms appear to be a primitive, non-specific response of uncertain significance but they may be responsible for immunopathological damage and, possibly, the initiation of auto-immune disease.

Congenital T-cell deficiency (the DiGeorge syndrome) results from hypoplasia of the thymus and leads to particular susceptibility to infections by intracellular parasites. Severe combined immunodeficiencies (SCID) are characterized by defective development of both B and T cells. A known cause is due to deficiency in adenosine deaminase (ADA), resulting in the accumulation of nucleic acid metabolites that are toxic for lymphocytes. Newborn infants with SCID are very susceptible to infection and rarely survive beyond the first year; treatment of SCID by specific gene therapy with ADA has recently entered clinical trials.

## 5.3 Antigenic diversity

Many pathogens exist in antigenically different forms. Each type may be able to cause the same disease but different types may cause different diseases.

Historically, the phenomenon of antigenic diversity was first identified with the streptococci, a bacterial genus that was initially divided phenotypically into three groups on the basis of the appearance of zones of hemolysis surrounding colonies of these bacteria on blood agar plates. There are $\alpha$-hemolytic streptococci (narrow zone of hemolysis contains some nonlysed erythrocytes and often green discoloration based on pigment produced by breakdown of hemoglobin), $\beta$-hemolytic strepto-cocci (broad, clear zone of hemolysis, colorless and free of intact erythrocytes) and $\gamma$-streptococci (produce no hemolysis). The $\beta$-hemolytic streptococci were divided by Lancefield into serologically distinguishable groups (serotypes) based on carbohydrate antigens obtained by acid extraction of the bacterial cell wall. Groups A–T have since been identified, and immunity to one serotype does not protect against any other serotype. Similar antigenic diversity is effected by the capsules of pneumococci; there are more than 80 different serotypes due to the chemical differences between polysaccharides of each serotype. Six types (a–f) of *H. influenzae* can be distinguished based on their capsular polysaccharides.

The antigenic typing of *Enterobacteriaceae* is also based on their surface antigens – H or flagella antigens; O or lipopolysaccharide surface antigens; K or capsule antigens; and Vi or virulence antigens, particularly of *S. typhi*. There are several hundred antigenically distinguishable types of *E. coli* but they are not all pathogenic, although particular O serotypes do appear to be associated with different clinical syndromes. The genus *Shigella* is divided into four groups on the basis of biochemical and serological tests – A (10 serotypes), B (six serotypes), C (15 serotypes) and D (one serotype). The genus *Salmonella* can be divided into over 1500 different serotypes.

Several viruses (measles, mumps, rubella) exist as one serotype but others show diversity ranging from three poliovirus serotypes and rotaviruses types A–D up to more than 100 different rhinovirus (common cold virus) serotypes.

## 5.4 Antigenic variation

The ability of a pathogen to vary its antigens either during or between infections enables the occurrence of persistent or recurrent infections.

*Neisseria* adhere to host cells via their pili, a process that can be blocked by specific antibodies. However, although there is an amino-terminal constant region of about 50 amino acids, there is variability in the pilin protein due mainly to sequences in the carboxyl-terminal part of the molecule. This is encoded in incomplete sequences which are recombined to give a complete gene product that may or may not be capable of assembly to give a new pilin molecule. Such recombinations occur with a frequency of $10^{-3}$ per cell division, giving rise to antigenic change.

Gonococci also have an external protein II or so-called opacity-associated protein (OPA) – this protein causes growing cells to adhere, making colonies opaque to transmitted light. There are 12 different OPA genes in *N. gonorrhoea* and three or four in *N. meningitidis*. Protein II genes have large numbers of CTCTT repeats immediately after the initiation codon but the number of repeats is varied so that the initiation codon is either in or out of frame. Consequently, translation occurs only when the appropriate number of repeats are present, and this happens at a frequency of about $10^{-3}$ per cell division. At any time the gonococcus may express from zero to three OPA. Meningococci also express highly variable outer membrane proteins.

*Trypanosoma brucei* has a 'variant surface glycoprotein' (VSG), and there are approximately 1000 VSG genes encoded by each organism. They may be switched randomly to produce new antigenic variants during the course of an infection.

Influenza virus shows antigenic changes that are effected by two different mechanisms. During this century there have been influenza pandemics in 1918, when 25 million people died, and less serious pandemics in 1933, 1946, 1957, 1968 and 1977 (*Figure 5.4*). Immune protection against influenza is provided mainly by antibodies against the virion envelope glycoprotein, the HA antigen, although some protection is afforded also by humoral responses to the NA antigen. To date, 14 different HA antigens and nine different NA antigens have been described but these are mostly from avian influenza viruses. Each human influenza pandemic has been marked by change or changes in one or both viral surface glycoproteins – 'antigenic shift'. This process occurs by genetic reassortment and the influenza virus genome is well-suited for this purpose. It consists of eight separate segments of RNA which facilitates their ready exchange in cells co-infected by two influenza viruses with different HA and NA antigens. Influenza A viruses occur very widely in nature, also infecting domesticated animals such as ducks and pigs. These

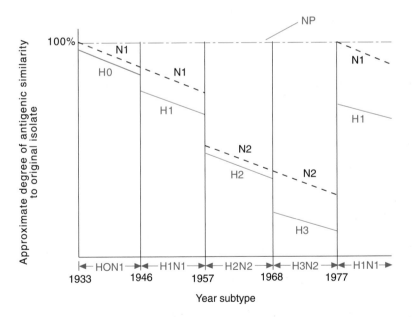

**Figure 5.4:** Antigenic shift and antigenic drift of influenza group A viruses. Antigenic shift is shown by the vertical lines and antigenic drift by the sloping lines. H, hemagglutinin; N, neuraminidase; NP, nucleoprotein. Derived from ref. [1] with permission from Blackwell Science Ltd.

animals seem to be the source of new influenza viruses, produced by genetic reassortment, to be introduced into the human population with resultant pandemic spread. More subtle antigenic changes in the influenza virus HA and NA glycoproteins can also occur by mutation – 'antigenic drift' – but this usually results in only local epidemics.

Antigenic shift giving rise to pandemics is seen only with influenza group A viruses. The HA and NA antigens are type-specific, and the neutralizing antibodies they elicit are effective only against the virions of a particular pandemic strain. However, the nucleoprotein (NP) which surrounds the genome is a group-specific antigen. Thus, pandemic influenza A viruses have different type-specific HA and/or NA antigens but they all share the same group-specific NP antigen. This antigen is inaccessible in the virion so it does not elicit a neutralizing antibody, but NP peptides are present on the surface of influenza virus-infected cells due to endogenous processing of viral proteins and their presentation on class I MHC molecules. Consequently, although protective humoral responses to influenza virus infection are restricted to one particular influenza virus A strain, cell-mediated immune responses do not have such limitations because they can recognize the group-specific NP antigen common to all influenza A viruses. Unfortunately from the viewpoint of vaccine design, such CTL appear to be more important in recovery from, rather than protection against, influenza virus infection.

Another molecular mechanism is responsible for the extensive antigenic variation seen with so many RNA viruses. During their replication, transcriptional errors are not corrected because there are no proofreading nucleases to remove mismatched bases from nascent RNA strands. The level of error in RNA transcription may, therefore, be 10 000-fold greater than in DNA replication. Such inherent genetic instability also gives rise to the high mutation rate seen with HIV. Complementary DNA is copied by reverse transcriptase from the retrovirus genome, again without correction of transcriptional errors, and further recombination may occur because the HIV genome is diploid. Antigenically different isolates of HIV can be recovered simultaneously from one patient and at different times during the course of infection from the same patient (*Figure 5.5*).

## 5.5 Antigenic complexity

Both protozoan and metazoan parasites show considerable variety of form in their life cycles with different antigens presented at each stage. In malaria, immunity to sporozoites gives protection against infection initiated by a bite from an infected insect but does not protect against merozoites, the blood stage of infection. Neutralizing antibodies and $CD8^+$ T cells confer protection against sporozoites but $CD4^+$ T cells aid the humoral defence against merozoites. Such antigenic complexity, together with the large number of irrelevant, decoy antigens and diversity of immune responses, overloads the immune system. It is, perhaps,

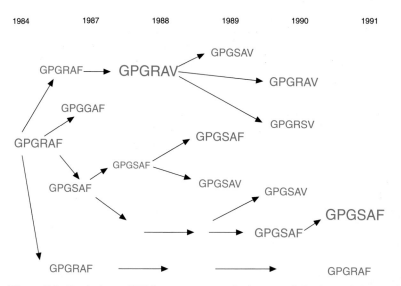

**Figure 5.5:** Evolution of V3 loop sequences during HIV infection. Phylogenetic relationships among V3 sequences obtained from seven time-points during an infection of an asymptomatic hemophiliac. The sequence of the hexapeptide at the 'crown' of the V3 loop is shown for each main lineage; lineages were also distinct at several flanking sites. The size of the lettering indicates the abundance of the genotype in each year. Reproduced from ref. [2] with permission from Current Science.

unsurprising that in endemic areas immunity to malaria develops only after repeated cycles of infection.

## 5.6 Latent infections

Primary infections with HSV type 2 may result in virus replication accompanied by clinical signs due to the damage caused to target organs, but this is followed by latent infection of neurones. The virus is transported up sensory nerve endings to dorsal root ganglia where the replication cycle stops at a very early stage with very limited expression of viral genes and no production of infectious progeny virions. This quiescent existence is also helped by the residence of latent viruses in a site that is not readily accessible to immune surveillance. Such latent infection may last for a few weeks before reactivation occurs, the virus travels back down nerves to cause recurrent lesions. The molecular mechanisms for reactivation are not well understood but various stimuli including fever, hormonal changes and immunodeficiency can act as triggers. Very severe genital herpesvirus infections can occur in immunocompromised individuals such as AIDS patients.

EBV, another herpesvirus, can also establish latent infections. In infectious mononucleosis, the infected B cells are polyclonally activated, and large amounts of heterophile antibodies are produced.

HIV can establish persistent infection by a different molecular mechanism. Immediately after entry into a host cell, reverse transcription of its diploid RNA genome is followed by integration of the proviral cDNA into host cell chromosomes. HIV encodes various regulatory proteins including a negative regulatory factor, *nef*, which inhibits proviral cDNA transcription and delays HIV replication. Such latent infection may persist for several years before further replication occurs and the infected individual progresses towards AIDS.

Persistent infection by measles virus appears paradoxically to be mediated by acquired humoral immune responses. Virus-specific antigens in the infected cell membrane react with a specific antibody, and the resulting immune complexes are shed. This phenomenon leads to the complete removal of measles virus-specific antigens from the surface of infected cells – 'antigen stripping' – and also results in downregulation of the synthesis of virus-specific polypeptides. Consequently, the measles virus-infected cell becomes unrecognizable by immune surveillance mechanisms. Subsequent reactivation of the persistent measles virus infection at some later time leads to SSPE.

## 5.7 Modulation of immune responses

In certain diseases, for example, leishmaniasis, production of Th1 cells favors recovery whereas Th2 cells exacerbate the disease. The elevated cAMP levels induced by cholera toxin result in inhibition of the proliferation of Th1 cells but have no effect on Th2 cells. There is also evidence that the Th1 subset of $CD4^+$ cells are specifically depleted during HIV infection.

## 5.8 Immune evasion

A particular feature of most protozoan and metazoan parasitic infections is their chronic nature. This ability to persist within the host is achieved by various mechanisms, and several have been described above. However, some parasites can actively avoid detection by host immune responses by their envelopment in a cloak of invisibility. Schistosomes *in vivo* are covered with host antigens – blood group determinants and MHC antigens – which give the parasite a coat of host molecules leading to recognition as self by the immune system. The significance of such 'antigen masking' is uncertain because these helminths can be grown in medium devoid of host molecules yet they remain resistant to attack by antibody and complement, possibly due to an inert outer tegument.

A number of intracellular parasites, particularly viruses, are able to evade immune recognition by CTL through the downregulation of MHC antigen expression by infected cells. A herpesvirus, HCMV, actually produces an MHC class I homolog that binds $\beta_2$-microglobulin although it is not known to interfere directly with CTL recognition. Other pathogens abrogate the effect of T-cell growth factor (IL-2): *T. cruzi* can downregulate receptors for IL-2; soluble IL-2 receptors appear during malaria; and *Pseudomonas* spp. produce proteases that cleave both IL-2 and IFN-$\gamma$.

HIV has an ultimate, but self-destructive, mechanism for evasion of the immune system. This is achieved through infection of phagocytic cells and CD4$^+$ Th lymphocytes followed by their subsequent lytic destruction. The consequent functional impairment of humoral and cellular immune responses enables unrestricted replication of this pathogen but its victory is short-lived. Acquired immunodeficiency resulting from HIV infection inevitably results in the acquisition of other infections by both opportunistic and pathogenic microorganisms. Inexorably, this leads to the death of the host.

## 5.9 Anamnestic immune responses

Immune responses are not forgotten but have the property of anamnesis, that is recollection. The primary immune response takes up to 14 days to become effective but it is accompanied by the generation of memory cells. These are almost immediately available to facilitate a secondary immune response that is more rapid, being effective within 1–2 days following antigenic stimulation (*Figure 5.6*). This is a particularly important principle of vaccination.

## 5.10 Protective immune responses

To design vaccines, it is important to know the nature of protective immune responses (*Table 5.1*). These examples apply mainly to acute diseases; those pathogens that cause persistent (latent or chronic) infections offer a much greater challenge.

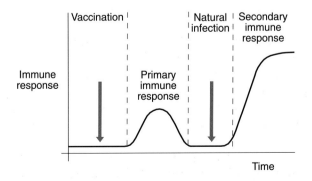

**Figure 5.6:** Primary and secondary immune responses.

**Table 5.1:** Immunity to infectious and parasitic diseases

| Disease | Pathogenesis | Major defence mechanism (target) |
|---|---|---|
| AIDS | Systemic infection | ?[a] |
| Cholera | Local infection Enterotoxin | IgA (toxin)[a] |
| Common cold | Local infection | IgA (virion) IFN, CMI (virus-infected cell) |
| Diphtheria | Local infection Exotoxin | IgG (toxin)[a] |
| Genital herpes | Local infection | ?[a] |
| Giardiasis | Local infection | IgA (parasite)? |
| Hepatitis B | Systemic infection | IgG (virion) IFN, CMI (virus-infected cell) |
| Influenza | Local infection | IFN, IgA, IgG (virion) CMI (virus-infected cell) |
| Leishmaniasis | Local/systemic | IFN-$\gamma$ and TNF enhance phagocytosis of parasite; IL-4 production worsens disease |
| Malaria | Systemic infection | ? IgG and CMI (circumsporozoite) |
| Measles | Systemic infection | CMI (virus-infected cell)[a] |
| Meningitis | Systemic infection | IgG (complement-mediated lysis and phagocytosis of bacterium)[a] |
| Pertussis | Local infection Exo- and endotoxins | IgG (toxins)[a] |
| Poliomyelitis | Systemic infection | IgG (virion)[a] |
| Typhoid | Systemic infection Endotoxin (LPS) | IgG (toxin and bacterium)[a] |
| Tuberculosis | Systemic infection | CMI (tubercles) |

[a]IgA protects at initial site of infection.
CMI, cell-mediated immunity.

# References

1. Taussig, M.J. (1984) *Processes in Pathology and Microbiology*. Blackwell Scientific Publications, Oxford.
2. Leigh Brown, A.J. (1991) *AIDS,* **5** (suppl. 2), S35.

Chapter 6

# Identification and analysis of vaccine antigens

## 6.1 The role of molecular biology in vaccine development

For a decade or more, molecular biology has been applied enthusiastically in the search for new vaccines to protect against communicable disease. Such intense interest in molecular technology has, however, tended to overshadow other important factors. Previous chapters have described the pathogenic mechanisms of infectious and parasitic organisms and the host's responses to infection. They show that, to design vaccines which will protect effectively, it is necessary to understand the biology of the etiological agents together with the immunology of the innate and induced responses they may provoke. This immunobiological knowledge may then be applied to better vaccine development at the molecular level.

A few fundamental questions must be answered before purposeful vaccine design can commence:

(1) How is the disease caused? How a pathogen replicates or multiplies is fundamental to an understanding of its pathogenicity, a qualitative or quantitative measure of its ability to cause disease, and its pathogenesis (those events that lead to clinical signs or symptoms of disease). An understanding of these phenomena is necessary in order to recognize targets for protective immune responses and to identify the particular body compartment(s) where they should be induced.

(2) What sort of immune response is required? Ideally, protective immunity should act as soon as possible after infection, requiring it to be present at, or close to, the pathogen's portal of entry; IgA at mucosal surfaces or IgG within the bloodstream. In many cases, however, cellular immune responses are also needed, particularly if disease is caused by an intracellular parasite. If a pathogen causes disease by the release of toxins, these soluble factors may be targeted by specific antibodies rather than the organism itself. It is clearly

important to ensure that these immune responses do not damage the host. Ideally, full qualitative and quantitative control should be maintained over the immune responses induced by a vaccine.

(3) How can protective immunity be produced? A vaccine based on the pathogen itself can be used safely only if it can be made available in a harmless form. The pathogen may be manipulated by physical, chemical or biological means that reduce or completely nullify its virulence. This enables the development of live, attenuated and killed vaccines, respectively, but their manufacture clearly requires strict safety control. Another potential disadvantage of these vaccines is their concomitant induction of irrelevant or even unwanted immune responses to other antigens. It may be more desirable, therefore, to use only that particular part of a pathogen which elicits protective immunity.

Molecular biology alone cannot answer all of these questions. The main aim of the succeeding chapters is to show how both immunological principles and molecular biological knowledge may be used together to create new vaccines. Methods for the identification of vaccine antigens are described, together with the application of new technologies for their expression and delivery, including recombinant live vectors. Molecular approaches to improve existing vaccines are also described.

## 6.2 Identification and cloning of antigens

A mysterious illness with high mortality appeared in the southwest United States in May and June 1993. Apparently healthy young adults aged between 20 and 40 years had an unexplained respiratory distress syndrome. It was suspected to be of infectious origin partly because of the quick onset, and 26 deaths were reported by the end of 1993 (see *Figure 6.1*).

The emergence of this new syndrome led to serious concern about its epidemic potential, and the power of molecular biology is well illustrated by the rapid scientific response. The first and vital clue came by screening sera from patients with the respiratory distress syndrome against a large number of pathogens. This demonstrated a low but significant level of serological cross-reactivity with hantavirus, or Hantaan virus, a rodent-borne RNA virus usually associated elsewhere in the world with hemorrhagic fever and renal disease. Complementary DNA (cDNA) sequences showed up to 70% homology with previously identified hantaviruses and lower but significant homologies with other bunyaviruses. It was decided, therefore, to try to amplify the genome of the unknown hantavirus and to determine its sequence by use of a reverse transcriptase polymerase chain reaction (RT-PCR) (see *Figure 6.2*). The cDNA sequences of known hantaviruses were aligned, and oligonucleotides synthesized, that represented the most conserved regions of the G2 glycoprotein. Using one pair of these conserved oligonucleotides, a fragment of cDNA was cloned by

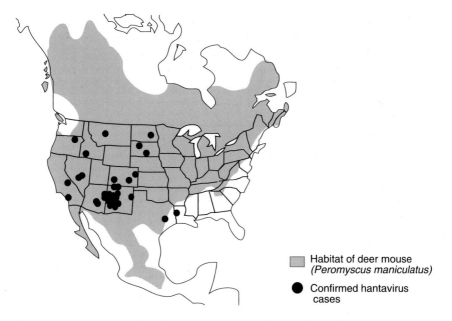

**Figure 6.1:** Location of confirmed pulmonary distress hantavirus infections. Orange areas indicate the extent of the deer mouse (*Peromyscus maniculatus*) habitat which appears to be the primary reservoir of the virus. Reprinted with permission from ref. [1]. ©1993 American Association for the Advancement of Science.

RT-PCR from diseased tissue taken at autopsy from a patient who died from the new mystery illness. Sequence data showed the cloned cDNA fragment to be part of the coding sequence of the G2 protein of the new hantavirus. Thus, the etiological agent of the new disease, and its major rodent vector, were identified within 30 days of the first death without even isolating the virus.

Further cloning and sequencing of the genome are likely to be the next steps in characterization of the virus, followed by expression of genes to provide antigens for diagnostic purposes or to make antisera.

Hantaan virus is approximately 90–100 nm in size and doughnut-shaped with a poorly defined arrangement of surface glycoproteins (*Figure 6.3*). Three unique negative-sense viral RNA species, L,M,S (large, middle and small), each form loosely helical configurations complexed with many copies of the nucleocapsid protein and a few copies of a large protein presumed to be the polymerase. L probably codes for the polymerase while the M segment codes for two glycoproteins G1 and G2 and the S segment codes for the nucleocapsid protein. This genomic structure and virion arrangement holds basically true for all bunyaviruses including Hantaan and is likely to be the case for the newly identified virus (now called pulmonary syndrome hantavirus, PSH).

Both G1 and G2 glycoproteins are present in equimolar proportions on the virion surface, and they are the main determinants of virulence. In

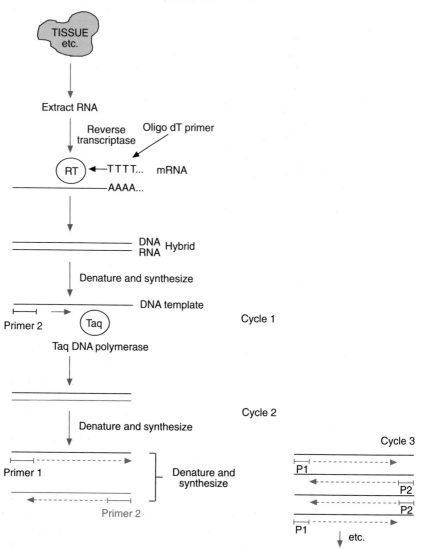

**Figure 6.2:** The basis of the reverse transcription polymerase chain reaction (RT-PCR). The synthesis and denaturation reactions are carried out in a single tube as Taq polymerase (or its equivalent) is highly thermostable. Primer 1 can be oligo-dT for priming all RNA species or a specific primer if only one mRNA species is to be amplified. Primer 2 provides the specificity if oligo-dT is used in the initial reverse transcription. Products are formed at an exponential rate. $2^n$ molecules of product are formed, where $n$ is the number of cycles used.

PSH they probably enable the virus to target the lung endothelial cells. Although an inactivated vaccine for hantavirus has been tested in humans in North and South Korea, more recent work towards a vaccine has been based on expression of G1 and G2. Monoclonal antibodies raised against

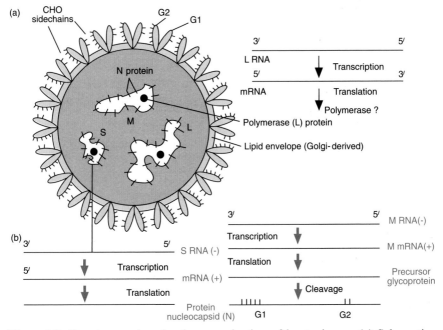

**Figure 6.3:** Structure and molecular organization of hantaviruses. (a) Schematic diagram of a hantavirus particle (80–120 nm). The RNA species are approximately 4.1 kb (L), 1.8 kb (M) and 0.863 kb (S) and complexed with N protein to form nucleocapsids. (b) Both the small and medium-sized RNAs are −ve sense in the nucleocapsids and are transcribed to give mRNA that can be translated into nucleocapsid protein(s) or the glycoproteins G1 and G2 (M). Although the strategy for production of the L protein has not been described, it is likely to be the same as for the nucleocapsid and glycoproteins.

these glycoproteins can neutralize virus *in vitro,* and vaccination with vaccinia recombinants (see Section 8.5) expressing either G1 or G2, or immunization with these glycoproteins purified from recombinant baculovirus-infected cells (see Section 7.2.4), can protect hamsters against virus challenge. Thus, it is possible that a vaccine to protect against the new respiratory distress syndrome could successfully be based on expression of the PSH G1 and G2 glycoproteins.

A molecular biological approach to vaccine development becomes more difficult as genomic complexity of the pathogen increases (*Table 6.1*). It has been known for many years that immunization with irradiated malarial sporozoites can protect against malaria. As the sporozoite stage in the life cycle of the malarial parasite can be grown only in small quantities, recombinant DNA technology has been used to identify, clone (see *Figure 6.4*) and express components of the sporozoite that might be of use in vaccine production. The genome of the malarial parasite is many thousand times larger than the genome of hantaviruses or HBV and therefore provides a different scale of problem. Not only were there little

**Table 6.1:** Genomic complexity of various pathogens

| Pathogen | Genome size[a] (bp) | Potential number of genes[c] |
|---|---|---|
| Hepatitis B virus | $3.2 \times 10^3$ (circular partially single-stranded DNA) | 7[d] |
| Influenza virus | $1.4 \times 10^4$ (eight segments double-stranded RNA) | 10[d] |
| Herpes simplex virus | $1.2 \times 10^5$ (linear double-stranded DNA) | $1.3 \times 10^2$ |
| *E. coli* | $4.2 \times 10^6$ (circular) | $3.5 \times 10^3$ |
| *S. cerevisiae* | $1.4 \times 10^7$ (16 pairs of chromosomes) | $1.2 \times 10^4$ |
| *Plasmodium falciparum* | $3 \times 10^7$ (14 chromosomes) | $2.5 \times 10^4$ |
| Parasitic helminths | $3 \times 10^8$ | $2.5 \times 10^5$ |
| Human cell | $2.9 \times 10^9$ (23 pairs of chromosomes[b]) | $2.4 \times 10^6$ |

[a]Genomes of HBV, influenza virus and HSV completely sequenced; *E. coli* close to being completed; yeast within 2 years.
[b]A single human chromosome is larger than the entire *P. falciparum* genome.
[c]Assuming $1.2 \times 10^3$ bp for an average gene.
[d]Actual number of gene products (some multifunctional).

sequence data available, there was also no idea of which gene products might have been protective. See Section 6.2.3 for detail of the initial cloning of the major malarial sporozoite surface antigen.

The starting point of any recombinant DNA work is to generate a library of DNA, in *E. coli*, which is representative of the pathogenic organism under study. For EBV, genomic DNA was isolated from purified virus particles and cleaved with two different restriction enzymes (*Bam*HI or *Eco*RI) before the fragments were cloned into the plasmid vector pBR322 in *E. coli* (for genomic map see *Figure 9.9*). Thirty-four *Bam*HI and 10 *Eco*RI fragments overlapped each other giving cloned DNA representative of the parent virus DNA. However, several of the virus DNA fragments were too large ($> 30$ kb) to be cloned in plasmid vectors such as pBR322 and cosmid vectors were used to transfer these fragments to *E. coli*. Thus, approximately 50 individual clones are representative of the 180 kb genome. For genomic libraries, phage λ or cosmids are used because they can accommodate between 20 and 40 kb of DNA requiring many fewer clones than plasmids to give complete libraries. The *E. coli* genome is approximately $4.2 \times 10^6$ kb and requires around 1000 clones with 20 kb (i.e. phage λ) inserts or 500 clones with 40 kb (i.e. cosmid) inserts to be truly representative. These could conceivably be stored as individual clones but, for more complex genomes, pools of clones are used, and these are referred to as libraries. The human genome has $2.9 \times 10^9$ bp and requires $7.6 \times 10^5$ clones with 20 kb inserts for a 99% probability that any one genomic fragment is represented. For cDNA libraries, many fewer clones will be required to give the same probability of the library containing any particular gene, due to noncoding repetitive cDNA present in such libraries. Techniques are available for screening the library to identify and isolate any gene of particular interest.

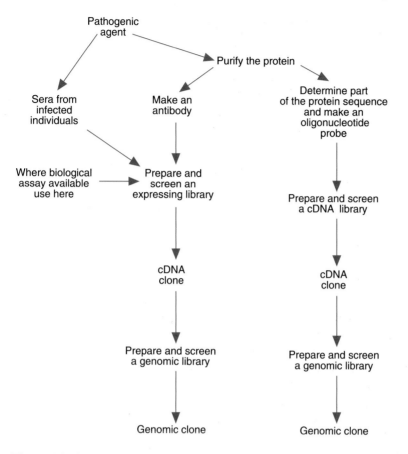

**Figure 6.4:** Common gene cloning strategies.

### 6.2.1 DNA/oligonucleotide hybridization

If sequence data of the pathogen's nucleic acid or purified mRNA is available, it is possible to detect recombinant clones by hybridization of $^{32}$P-labeled DNA or RNA to bacterial colonies or bacteriophage plaques. Often a protein has been purified and sufficient amino acid sequence is available to allow a corresponding nucleic acid sequence to be synthesized. Due to the degeneracy of the genetic code, a complex mixture of oligonucleotides is required to ensure that all possible sequences are represented. Labeling this mixture of oligonucleotides yields a probe that can be used to screen a genomic (or cDNA) library that might be expected to contain the gene(s) of interest (see *Figure 6.5*).

### 6.2.2 Hybrid selection and cell-free translation

This approach was used to map genes of large DNA viruses but has now been largely superseded by expression cloning. It is possible to use hybrid

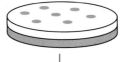

Agar plate with bacterial
colonies or phage plaques

Transfer the bacterial colonies or phage plaques
to filter and lyse to release DNA

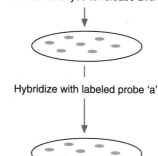

Hybridize with labeled probe 'a'

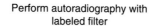

Perform autoradiography with
labeled filter

Phage plaque or
bacterial colony ⟶
containing nucleic
acid 'a'

**Figure 6.5:** Screening clones by *in situ* hybridization. This technique is equally applicable to phage plaques or bacterial colonies. When screening a library, the density of plaques or phage is very much higher than shown here, with perhaps 100 000 colonies or plaques on a 15-cm diameter plate. After screening, a plug of agar is lifted out and the bacteria or phage eluted and replated at lower density. This process is repeated until a pure bacterial clone or phage population is obtained. Probe 'a' can be labeled RNA, DNA or oligonucleotide. In the case of oligonucleotides, problems can arise from the degeneracy of the genetic code. For example, if a protein contained the amino acid sequence 'Phe, Trp, Gly, Leu, Leu, Tyr' then, in order to detect all possible coding sequences, it would be necessary to synthesize the multiply degenerate oligonucleotide:

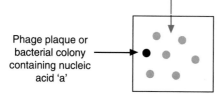

i.e. a mixture of 1024 oligonucleotides. Of these alternatives, only one will be the authentic coding sequence. Reproduced from Williams, J. *et al.* (1993) Genetic Engineering, BIOS Scientific Publishers, Oxford.

selection of mRNA coupled with cell-free translation to map protein-coding sequences. DNA clones from a library, either individually or in pools, can be immobilized on a solid support and mRNA added. Only the mRNA that hybridizes specifically to the clone(s) will bind, and this can be eluted and translated in a cell-free system. The protein can be identified by immunoprecipitation or assayed for appropriate activity (see *Figure 6.6*).

An example that encompasses both this strategy and the sequence route involves the identification of the major membrane antigen gene, gp340/220 of EBV. In about 1980, the EBV genome was cloned and, in 1984, the entire sequence of EBV DNA was published. Using computer programs, the gp340/220 gene was predicted to be an open reading frame (ORF) in the *Bam*HI L fragment. This was confirmed a year later by hybrid selection of EBV mRNA to DNA clones of *Bam*HI L followed by cell-free translation of the eluted mRNA and immunoprecipitation of gp340/220

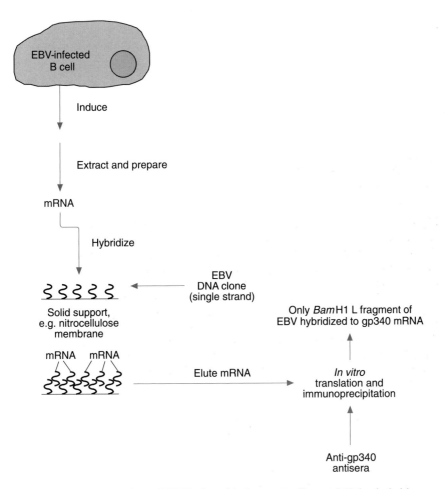

**Figure 6.6:** Identification of EBV plasmid clone encoding gp340 by hybrid selection and translation.

with a high titer antibody. The ORF that hybridized with the gp340/220 mRNA was the one predicted to encode the gp340/220 gene by computer analysis. The hybrid selection approach is rather labor intensive and has, for the most part, been superseded by one of the forms of expression cloning.

### 6.2.3 Expression cloning

This approach is invaluable when the only means of identification is an antiserum against the protein or pathogen of interest. It can also be used to identify clones that code for immunogenic proteins of a pathogen. Antisera to the pathogen are used to screen expression libraries; those clones that bind the sera will encode immunogenic proteins (*Figure 6.7*).

Probably the most laborious form of this approach is its use in conjunction with a biological assay. cDNA libraries are cloned from cells where the desired gene is known to be expressed and inserted into a plasmid that will allow

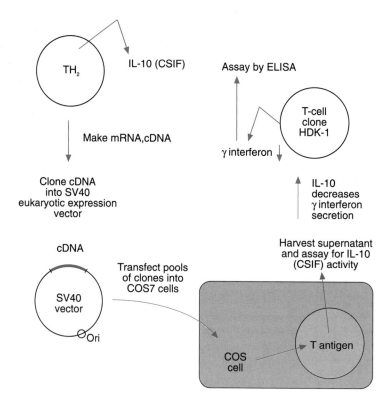

**Figure 6.7:** Cloning of IL-10 (cytokine synthesis inhibitory factor, CSIF) using transfection and biological assay. Pools of cDNA clones from an IL-10 expressing cell line in an SV40 vector were transfected into COS7 cells. The vector replicates in COS cells and pools of clones expressing IL-10 were identified. Subsets of the positive pools were screened until eventually a single clone expressing murine IL-10 was identified.

expression in eukaryotic cells. Clones or pools of clones are then transferred to appropriate cell types and either cell extracts or cell supernatant assayed for biological activity. If a pool of clones gives the biological activity, individual clones can be transferred and assayed to identify the desired cDNA clone. This methodology, although tedious, has allowed many of the interleukin genes to be cloned, for example IL-3 and IL-10, probably because the biological assays for these proteins are very sensitive. Identification of cDNA encoding cell surface proteins can be carried out in an analogous way. Once again, plasmid cDNA pools are prepared and transfected into the recipient cell. In this case, a radiolabeled or fluorescent receptor ligand is added and allowed to bind to the cells. After washing, retention of the tagged ligand indicates a positive pool of cDNAs. Subpopulations of the pool are made and re-assayed until a single cDNA clone encoding the gene of interest is isolated.

Other gene products or vaccine antigens may require an enrichment step. Many genes expressed on the cell surface, such as receptors and adhesion molecules, have been cloned by 'panning' techniques (see *Figure 6.8*). Here the cells expressing the required gene are selected out with antibody, immobilized ligands or by interaction with other cells. cDNA libraries are constructed using a eukaryotic expression vector and cloned into *E. coli*. The library is transferred to cultured cells, resulting in cells expressing a vast array of genes with the gene to be cloned expressed in a limited number of cells. These cells can be enriched relative to nonexpressing cells by 'panning' the transfected cells on plates containing an antibody to the

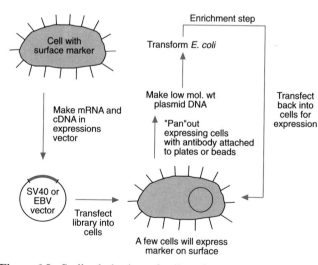

**Figure 6.8:** Stylized cloning of cell surface markers using 'panning' techniques. Enrichment of positive clones is achieved by selecting cells that express the desired protein with antibody. Vector DNA is prepared from these cells and transferred back to *E. coli*. The *E. coli* library, or subsets of the library, can then be tested for ability to bind antibody or subjected to a further round of enrichment. Reproduced from Williams, J. (1993) *Genetic Engineering,* BIOS Scientific Publishers, Oxford.

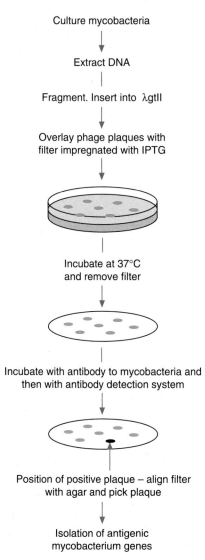

Culture mycobacteria

Extract DNA

Fragment. Insert into λgtll

Overlay phage plaques with
filter impregnated with IPTG

Incubate at 37°C
and remove filter

Incubate with antibody to mycobacteria and
then with antibody detection system

Position of positive plaque – align filter
with agar and pick plaque

Isolation of antigenic
mycobacterium genes

**Figure 6.9:** Screening lambda expression libraries. λgt11 contains the *lac*Z promoter so gene expression is induced with the galactoside IPTG. The induced protein, released from the lysed cells, then sticks to the filter.

gene of interest. Cells expressing this gene will bind preferentially. Alternatively, antibody or ligand can be attached to magnetic beads, and positive cells selected with a magnet. Eukaryotic expression plasmid DNA is made from the cells that bind, and transferred back into *E. coli*. This cycle can be repeated until eventually pure clones are isolated.

The most extensively applied form of expression cloning involves the use of plasmid or bacteriophage vectors in *E. coli* and identification of DNA clones using antisera to the gene product. Here a vector such as the bacteriophage λgt11 is constructed so that cDNA fragments can be cloned into sites adjacent to the β-galactosidase gene (see *Figure 6.9*). The recombinant λ phage-infected bacteria express a β-galactosidase fusion protein containing epitopes present in the cDNA. Recombinant phage are detected with monoclonal antibodies, monospecific antisera or even polyclonal antisera with many antibody specificities present. The cDNA insert is then sequenced, and the whole gene isolated in a conventional way.

A variation on this method allowed the initial cloning of the malarial sporozoite surface antigen. Malarial sporozoite stage cDNAs were introduced into the ampicillin resistance gene (*amp*) of the plasmid pBR322. Low levels of expression of the sporozoite surface antigen as fusions with the *amp* gene were detected by solid-phase radioimmunoassay using a monoclonal antibody specific for the protein. In this way a cDNA clone coding for the antigen was isolated and subsequently sequenced.

## 6.3 Characterization of vaccine antigens: B-cell epitopes

### 6.3.1 Computer-aided prediction of B-cell epitopes

Once a gene coding for a potential vaccine antigen has been cloned, the next step in its characterization is to determine the DNA sequence of the gene. From these sequences it is a relatively simple step to determine the amino acid sequence of the encoded protein. Virtually all genes start with an AUG triplet encoding methionine, simplifying the identification of the translational initiation site. *Table 6.2* shows the context around AUG triplets known to initiate protein translation. A distinct pattern, known as Kozak's rule, can be seen with −3 and +4 being the most important nucleotides in determining efficient translational initiation. Other information can be gleaned from genomic DNA clones. Motifs important in transcriptional initiation, in the binding of transcriptional factors, in termination and in polyadenylation signals have all been described. These data give important corroborative evidence that the gene under investigation is functional.

Armed with the likely translational start, the amino acids coded by triplets in the sequence can be decoded, translation terminating with TAA, TGA or TAG. Occasional complications in determining the

**Table 6.2:** Nucleotide sequences flanking eukaryotic start codons. Combined figures for vertebrates and vertebrate viruses are shown ($n = 2944$)

| Nucleotide | Position relative to start codon | | | | | | | | | | | | | | |
|---|---|---|---|---|---|---|---|---|---|---|---|---|---|---|---|
| | −10 | −9 | −8 | −7 | −6 | −5 | −4 | −3 | −2 | −1 | +1 | +2 | +3 | +4 | +5 |
| A | 28 | 25 | 23 | 25 | 23 | 22 | 25 | 57 | 34 | 25 | | | | 24 | 31 |
| G | 22 | 28 | 23 | 21 | 36 | 20 | 18 | 30 | 14 | 22 | | | | 50 | 20 |
| C | 29 | 23 | 34 | 28 | 20 | 34 | 43 | 7 | 37 | 43 | | | | 12 | 35 |
| U | 22 | 23 | 22 | 25 | 23 | 24 | 13 | 6 | 16 | 10 | | | | 13 | 14 |
| Consensus | c/a | g | c | c | g | c | g | A/G | a/c | c | AUG | | | G | a/c |

amino acid sequence of a protein can arise from phenomena such as translational frame shifting, ribosomes 'jumping' over untranslated bases or post-transcriptional RNA editing. However, these problems do not affect the vast majority of genes. Other features of a protein that can be discerned from its primary amino acid sequence are:

(1) potential N-linked glycosylation sites (N-linked glycosylation occurs at the amino group of asparagine in the motif asparagine, $X$ (anything), serine or threonine);
(2) fatty acids can be added to the 5′ end of molecules; in particular MG$XXX$[S/T/A] (where M,G,$X$,S,T,A are methionine, glycine, any amino acid, serine, threonine and alanine, respectively) indicates that a myristoyl group may have been added;
(3) 18–21 amino acid hydrophobic 5′ sequences are indicative of a signal sequence which is inserted across the endoplasmic reticulum and cleaved during transit of a protein to the cell surface; and
(4) internal hydrophobic sequences can be indicative of regions of the molecule that interact with membranes. In a type 1 glycoprotein with a signal sequence, an internal hydrophobic domain 20 or more amino acids long (possibly towards the C terminus of the protein) may be the membrane anchor of the protein.

Many other motifs which may have significance have been observed, for example consensuses for protease cleavage sites, phosphorylation sites and ribosomal frameshifts.

Immunogenic regions of proteins can be formed from contiguous amino acid sequences (so-called 'linear epitopes') or from noncontiguous amino acids brought together by protein folding (so-called 'conformational epitopes') (see *Figure 6.10*). Secondary structure prediction is currently imprecise, and it is not possible to predict conformational epitopes. The prediction of linear epitopes has been attempted by the use of a variety of computer programs. Those in refs [2] and [3] sum the hydrophobicity or hydrophilicity of particular amino acids for between eight and 12 residues and plot the result as a running score. Residues with the greatest hydrophilicity scores are considered to be the most likely to be antigenic. Other programs attempt to calculate the surface probability or flexibility of a group of amino acids on the assumption that if an epitope is flexible or predicted to be on the surface, it is more likely to be an antigenic epitope.

Secondary structure prediction algorithms have also been developed and have often been used to predict secondary structure when the quantity of protein required for the physical techniques associated with structural analysis is not available. The best known are the algorithms in refs [4] and [5]. Chou and Fasman analyzed crystallographic structures of 15 proteins and calculated the likelihood of a particular amino acid being present in α-helix, β-sheet, random coil and reverse turn. Predictions of the conformation of an unknown protein are then made by calculating the probabilities of the amino acids in the unknown protein folding into one of

(a)

| Approach | Linear | Conformational | Post-translational |
|---|---|---|---|
| Prediction from amino acid sequence | + | + | + |
| Prediction from protein secondary structure | + | + | + |
| Western blot (whole protein or fragment) | + + + | + | − to + [a] |
| ELISA (whole protein or fragment) | + + + | + + | − to + + [a] |
| Immunoprecipitation (whole protein or fragment) | + + + | + + | − to + + [a] |
| Pepscan (based on peptide reactivity) | + + + | − | − |
| Crystallography of antigen–antibody complex | + + + | + + + | + + + |

[a]Value depends on the system used for expression and whether post-translational modification occurs.

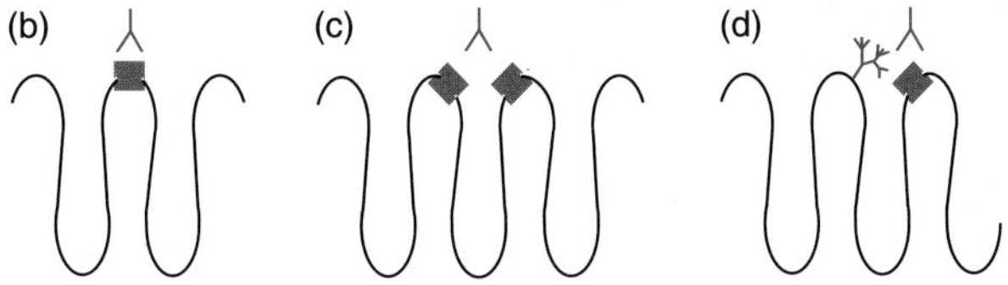

**Figure 6.10:** (a) Relative value of various techniques in determining B-cell epitopes. The structure of B-cell epitopes varies widely, from linear epitopes (b) consisting of a string of amino acids to conformational epitopes (c) that are formed from nonadjacent amino acids as a result of protein folding. Post-translational epitopes (d) can arise from recognition of structures such as carbohydrate or as a result of structural changes in a molecule due to post-translational modification.

these structures. There are basic limitations to such analysis: a limited number of crystal structures have been used; no account was taken of the effect of neighboring amino acids (although the algorithm of Garnier *et al.* [5] does allow for this); the difficulties of taking into account post-translational modifications such as glycosylation; and contacts made with other proteins (or as dimers, trimers or other polymers). Secondary structure predictions for HIV-1 glycoprotein gp160 are shown as a two-dimensional 'squiggles' plot in *Figure 6.11*, with antigenic index superimposed.

Jamieson and Wolf (see ref. [6]) have attempted to give a more reliable indication of antigenicity by weighting parameters determined by other programs and giving a score termed the 'antigenic index'. *Figure 6.12* shows a compilation of the known antigenic sites of HIV-1 gp160. The success (or otherwise!) of computer prediction of B-cell epitopes is evident from a comparison with *Figure 6.10*.

It is widely acknowledged that prediction of antigenic epitopes is unreliable even with the knowledge of many more crystal structures and

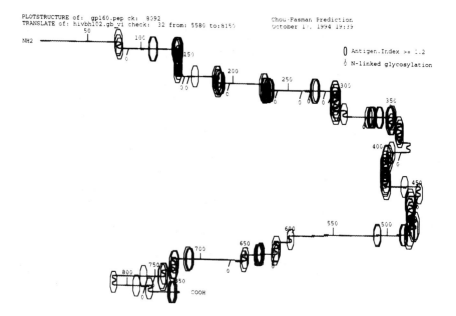

PLOTSTRUCTURE of: gp160.pep ck: 8392
TRANSLATE of: hivbhl02.gb_vi check: 32 from: 5580 to:8159
Chou-Fasman Prediction.
October 1, 1994 19:39

0 Antigen.Index >= 1.2
δ N-linked glycosylation

**Figure 6.11:** Two-dimensional 'squiggles' plot of secondary structure with the 'antigenic index' of Jamieson and Wolf [6] superimposed. Printout from the Wisconsin sequence analysis package reproduced with permission from Genetics Computer Group Inc.

the availability of more sophisticated algorithms. Most researchers apply all the methods available and test the conclusions experimentally. With the current rate of improvements in computer graphics, in structure modeling and an ever-expanding database of crystal structures, the full potential of protein secondary structure prediction may well become more apparent.

### 6.3.2 Comparative nucleotide sequence analysis

If a large number of different strains or isolates of an organism is available, it may be possible to identify immunogenic B-cell epitopes from DNA sequence data by comparison of deduced amino acid sequences. For example, the virion protein VP1 of poliovirus type 3 showed a highly variable region between amino acids 98 and 105 which was predicted to be the major antigenic site. This prediction was based on the observation in a variety of systems that those amino acids which change most frequently are under immune pressure and reside on the external surface of the pathogen. Similarly, sequencing studies on the hypervariable regions of HIV-1 gp160 not only indicate the probable antigenic sites, they also provide data on the relatedness of virus strains.

This method has been most successful with virus isolates where predictions can be made across serotypes, as with influenza virus

(a)

| Region | Residues | Sequence |
|---|---|---|
| C1 | 102–126 | MWKNDMVEQMHKIISLWDQSLKPC |
| V2 | 164–187 | CSFNISTSIRGKVQKEYAFFYKLD |
| C2 | 254–274 | CTHGIRPVVSTQLLLNGSLAE |
| C2/V3 | 298–314 | SVEINCTRPNNNTRKSI |
| C2/V3 | 303–338 | CTRPNNNTRKSIRIQRGPGRAFVTIGKIGNMRQAHC |
| C2/V3 | 303–321 | CTRPNNNTRKSIRIQRGPG |
| V3 | 306–338 | PNNNTRKSIRIQRGPGRAFVTIGKIGNMRQAHC |
| V3 | 308–331 | NNTRKSIRIQRGPGRAFVTIGKIG(C) |
| V3 | 314–328 | IRIQRGPGRAFVTIG |
| V3 | 361–392 | GNNKTIIFKQSSGGDPEIVTHSFNCGGEFFYC |
| C3 | 423–437 | IINMWQKVGKAMYAP |
| C3 | 425–452 | CRIKQIINMWQEVGKAMYAPPISGQIRC |
| V5 | 458–484 | GLLLTRDGGNSNNESEIFRPGGGDMRD |
| C4 | 477–508 | PGGGDMRDNWRSELYKYVVLIEPLGVAPTLA |
| C4/C5 | 503–532 | VAPTKAKRRVVQREKRAVGIGALFLGFLGA(G) |
| C5 | 518–542 | RAVGIGALFLGFLGAAGSTMGAASM |
| C5 | 579–605 | GIKQLQARILAVERYLKDQQLLGIWGC |
| C5 | 616–632 | PWNASWSNKSLEQIWNN |
| C6 | 735–752 | (Y)DRPEGIEEEGGERDRDRS(G)(C) |
| C6 | 735–752 | DRPEGIEEEGGERDRDRS |

Reproduced from ref. [7] with permission from the Federation of American Societies for Experimental Biology.

(b)

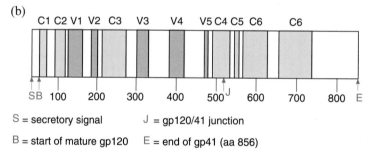

S = secretory signal  J = gp120/41 junction

B = start of mature gp120  E = end of gp41 (aa 856)

**Figure 6.12:** Summary of HIV neutralizing epitopes within HIVgp160. C and V represent constant and variable domains of gp160.

hemagglutinin H1, H2 and H3, whereas for other viruses that have only a single serotype, such as measles and mumps, this approach has been unrewarding.

### 6.3.3 Monoclonal antibody escape mutants

Cultivation of a virus in the presence of a neutralizing monoclonal antibody allows the selection of mutants that escape neutralization. DNA sequence analysis of these escape mutants allows identification of those parts of the protein that differ from wild-type, neutralized virus. Such sequences, therefore, must encode those amino acids that make the epitopes which bind neutralizing antibody. This type of analysis is not just limited to epitopes that raise neutralizing antibodies; amino acids involved in antibody binding can also be identified by including complement in the medium when selection takes place. Bound antibody that fixes complement will kill the infected cell and lead to the selection of viruses that cannot bind the antibody. Antibodies that recognize linear or conformational epitopes can be used but, in cases where secondary structures are recognized, sequence analysis may indicate other amino acids that are important in maintenance of conformation as well as those recognized by direct binding to the antibody.

In the case of influenza virus HA (see *Figure 6.15*), the epitopes recognized by neutralizing antibodies were identified both by sequence variation in natural isolates and by neutralization escape mutants.

### 6.3.4 Analysis of fragments of the antigen

The traditional approach to delineation of antigenic epitopes is to isolate the protein of interest, generate fragments by biochemical or enzymatic means and analyze the fragments for antibody binding. This strategy is dependent on the availability of large amounts of antigen which is particularly difficult with pathogens that cannot readily be grown *in vitro*.

Another way of generating specific fragments of the protein is to express the appropriate gene or parts of it in *E. coli* either as fusions with other proteins or on their own (*Figure 6.13*). A series of truncations increasing in size can be generated and assayed for monoclonal antibody binding. In the hypothetical example shown in *Figure 6.13* the protein encoded by clone 3 is recognized by antibody 1, while the protein encoded by clone 4 is not. From this it can be inferred that antibody 1 binds to the amino acids lost from protein 3 to generate protein 4 while antibody 2 binds to the amino acids lost between proteins 1 and 2. Lack of binding to a protein may mean loss of secondary structure rather than loss of the epitope. Thus this technique will determine linear epitopes but will not identify conformational epitopes.

A further method of producing fragments of an antigen which may bind antibodies differentially is to generate RNA *in vitro* and translate this

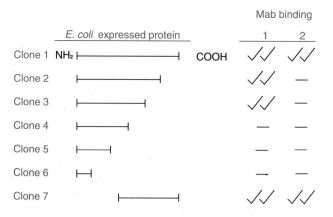

**Figure 6.13:** Mapping of monoclonal antibody reactivity with proteins expressed in *E. coli* (can be native or fused to other proteins).

RNA into protein. The gene of interest is cloned into a vector next to a promoter such as the phage T7φ10 promoter. Templates for the T7 RNA polymerase are generated by cleavage of the plasmid clone with several restriction enzymes. Restriction enzymes are chosen which truncate the gene at various points along its length. RNA generated from these templates is then translated into protein in an *in vitro* translation system and tested for antibody binding by immunoprecipitation. The linear epitope recognized by a monoclonal antibody is contained within the region lost by truncation that corresponds to a loss of antibody binding.

### 6.3.5 Epitope scanning

Once the amino acid sequence of a protein is known (or has been deduced from the DNA sequence), short overlapping peptides can be synthesized covering the whole length of the antigen sequence. These peptides, usually between eight and 10 amino acids in length, can then be probed in an enzyme-linked immunosorbent assay (ELISA) format by monoclonal or polyclonal antibodies raised either by immunization or by natural infection (*Figure 6.14*). Specific antibody binding indicates the antibody sequence of the epitope that is recognized. The major limitation for most laboratories is the expense of synthesizing hundreds of overlapping peptides even though peptides can now be synthesized on plastic pins in a format compatible with 96-well ELISA plates. Once an epitope has been identified, re-synthesis of peptides with single amino acid substitutions can reveal which amino acids are crucial in antibody recognition.

This technique is able to determine linear epitopes but will not identify conformational epitopes. Accurate antibody reactivity will also be difficult to demonstrate where post-translational modification of the primary protein translation product has occurred, for example, N-glycosylation at asparagine residues in the NXT/S motif.

### 6.3.6 Three-dimensional structure

The structure of several viruses and vaccine antigens has been determined by X-ray crystallography. Most notably, three-dimensional structures for the related picornaviruses poliovirus, human rhinovirus 14 and foot-and-mouth disease virus, and for H3N2 Hong Kong influenza virus HA and NA, have been described. The type of information gained by X-ray crystallography is illustrated by the structure of influenza virus HA shown in *Figure 6.15*. Several features become apparent when the positions of amino acid changes, identified in natural variants or by monoclonal antibody escape mutants, are superimposed on this structure. The major antigenic regions (A–E) are on the surface of the molecule, and only one is buried to any extent at the junctions of the trimers. The rest of the HA molecule, including the receptor site for sialic acid (which is in a pocket at the distal end) is more constant.

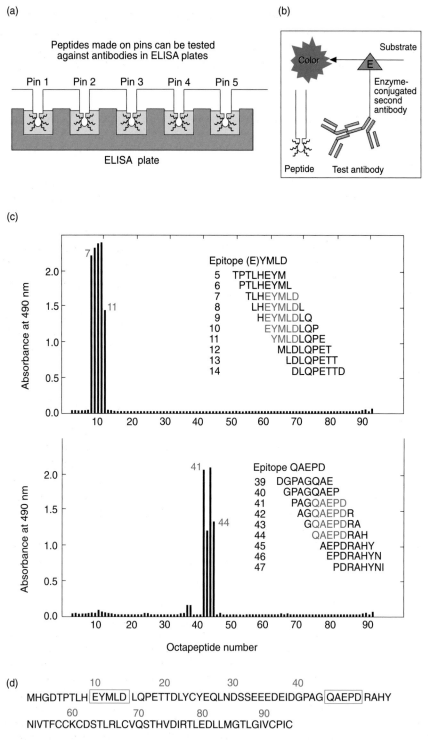

(a)

Peptides made on pins can be tested against antibodies in ELISA plates

Pin 1　Pin 2　Pin 3　Pin 4　Pin 5

ELISA plate

(b)

Substrate

Color

E

Enzyme-conjugated second antibody

Peptide　Test antibody

(c)

Absorbance at 490 nm

Epitope (E)YMLD

5　TPTLHEYM
6　PTLHEYML
7　TLHEYMLD
8　LHEYMLDL
9　HEYMLDLQ
10　EYMLDLQP
11　YMLDLQPE
12　MLDLQPET
13　LDLQPETT
14　DLQPETTD

Absorbance at 490 nm

Epitope QAEPD

39　DGPAGQAE
40　GPAGQAEP
41　PAGQAEPD
42　AGQAEPDR
43　GQAEPDRA
44　QAEPDRAH
45　AEPDRAHY
46　EPDRAHYN
47　PDRAHYNI

Octapeptide number

(d)

　　　　　10　　　　20　　　　30　　　　40
MHGDTPTLH EYMLD LQPETTDLYCYEQLNDSSEEEDEIDGPAG QAEPD RAHY
　　　60　　　　70　　　　80　　　　90
NIVTFCCKCDSTLRLCVQSTHVDIRTLEDLLMGTLGIVCPIC

Human papillomavirus type 16 E7 protein

The crystal structure of poliovirus capsids has revealed that the major antigenic reactivity within the capsid antigen VP1 can be mapped to loops that appear to project from the surface of the virus particle. This again emphasizes the importance of surface structures in the recognition of antigens by antibody.

## 6.4 Characterization of vaccine antigens: T-cell epitopes

As mentioned previously (Section 4.6), T lymphocytes recognize foreign antigens, processed into peptides, that are associated with the extracellular portion of the MHC molecule. Helper CD4$^+$ T cells recognize antigen in association with MHC class II molecules, whereas cytotoxic CD8$^+$ T cells recognize antigens in association with MHC class I molecules. The genetic polymorphism of MHC molecules determines the conformation of the cleft and, consequently, both the specificity and affinity of peptide binding in T-cell recognition. This means individuals with different MHC haplotypes can vary considerably in their ability to generate cell-mediated immune reactions against the same antigen.

It is clearly important to determine which epitopes (peptides) within a potential vaccine antigen or a pathogen will associate with a particular MHC molecule. It should be noted that not all peptides which bind to MHC class I or II molecules will generate CTL.

### 6.4.1 The use of vaccinia recombinant viruses and peptides

Vaccinia recombinants have proved to be invaluable in identifying the targets of CTL. This is due to the ability of the recombinant viruses to express the foreign antigen on the infected cell surface in conjunction with MHC molecules, as illustrated in *Figure 6.16*. These cells may be recognized and lysed by autologous CTL directed against the foreign antigen. Effector cells can either be bulk lymphocyte cultures or T-cell clones. The extent of CTL lysis is quantified by labeling the infected cells with $^{51}$Cr and measuring the amount of radioactivity released. Processing of the foreign antigen expressed by the recombinant virus

---

**Figure 6.14:** (Opposite) Epitope mapping of antibody reactivity. (a) Large numbers of antisera can be tested for binding by synthesizing peptides on pins that fit into a 96-well plate. Reproduced from Ollier, W. and Symmons, D.P.M. (1992) *Autoimmunity*, BIOS Scientific Publishers, Oxford. Antibody can be bound in small volumes to an array of peptides and reactivity detected by an enzyme-conjugated second antibody (b). (c) shows the reactivity of two different monoclonal antibodies to human papillomavirus type 16 E7 with octapeptides spanning the E7 protein. Each peptide is displaced by one amino acid from the previous one and is identified in the histogram by the number of the first amino acid in its sequence. Part (c) reproduced from ref. [8] with permission from the Society for General Microbiology.

**(a)**                                                    **(b)**

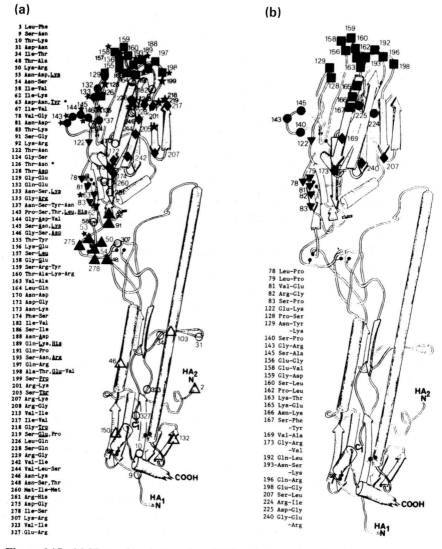

**Figure 6.15:** (a) Natural variation since 1968 and monoclonal variants suggest the antibody binding sites on the 1968 HA. ●, Site A; ■, site B; ▲, site C; ◆, site D; ▼, site E. The symbols represent locations of natural sequence variation between 1968 and 1979. Single-site monoclonally selected variant;

★ plus site symbol represents a site of natural variation that has also been observed in a monoclonally selected variant. Underlined amino acids in the list of amino acid substitutions were observed in monoclonally selected variants only; no underline indicates in natural variants only, first letter underlined indicates substitutions found in both natural and monoclonally selected variants. An asterisk indicates addition of an N-glycosylation site; a minus indicates the loss of such a site. Reproduced from ref. [9] with permission from Raven Press. (b) Monoclonal antibody-selected variants of the A/PR/8/34 H1 hemagglutinin shown on the 1968 H3 structure for comparison with the antigenic structure of the H3 strains in (a). Reproduced with permission from *Nature*, Colman *et al.* (1987). ©1987 Macmillan Magazines Limited.

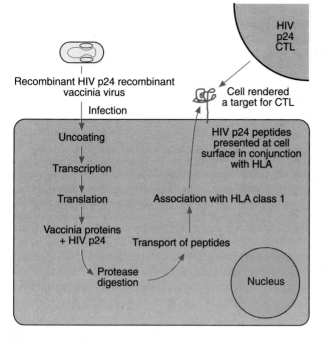

**Figure 6.16:** Generation of CTL targets for HIV p24 by infection with a vaccinia virus. A variety of different cells with different HLA types can be infected and used as targets.

appears to take place authentically. Interestingly, cells infected with a vaccinia recombinant expressing a 15 amino acid peptide from the nucleoprotein, an internal group antigen, of influenza A virus were lysed by CTL specific for the peptide sequence.

An advantage of this system is that vaccinia recombinant viruses can themselves prime and stimulate cell-mediated immune responses in vaccinated animals. The specificity of CTL induced by these vaccinations can be tested on cells infected with different strains and types of the appropriate target virus. For example, individual vaccinia recombinant viruses expressing each of the 10 genes of influenza A virus as well as several engineered genes have been constructed and used to analyze CTL responses in animals and humans. These recombinants have been used to show individuals previously infected with influenza A can recognize matrix M1, polymerase PB2 and nucleoprotein in conjunction with MHC class I antigens. Furthermore, transfection protocols and the vaccinia recombinant approach have indicated that influenza HA-specific CTL induced in animals are not usually cross-reactive but nucleoprotein is the major target for cross-reactive, influenza-specific CTL.

Vaccinia recombinant viruses which express truncated genes can also be used to generate targets for CTL. Thus, specific epitopes within a protein that generate T-cell reactivity can be identified. For example, the reverse transcriptase (RT) of HIV type 1, together with a number of truncations of RT, were expressed in vaccinia virus and used to create targets for anti-HIV CTL clones and bulk lymphocyte cultures. Two clones mapped to between amino acids 422 and 315, and five other clones studied mapped to between amino acids 531 and 480 of HIV-1 polymerase. The sequences recognized by the CTL were further mapped using peptides, corresponding to the polymerase gene.

### 6.4.2 In vitro *binding assays, CTL generation, HLA transgenics*

Presentation of peptides bound to an MHC class I molecule is a prerequisite for stimulation of CTL responses. One way of analyzing an antigen for potential CTL epitopes would be to synthesize a set of overlapping peptides that correspond to the amino acid sequence of the protein and test the ability of each peptide to bind to class I molecules. This has been achieved by isolating MHC class I molecules (e.g. HLA A2.1) and assessing binding in a competitive assay with a known labeled MHC class I 'binder'. *Figure 6.17* shows a summary of the HLA A-specific motifs as well as the effects of replacing one amino acid at a time on the binding of peptides to HLA A2.1. Similar assays have been done on other MHC class I molecules including HLA A1, A3, A11 and A24, A68, B8, B27 and B53, as well as with MHC class II molecules such as HLA DQ3.1 and numerous HLA DR alleles, showing that the motifs required for binding are quite different. MHC class II molecules have a less stringent requirement for binding.

Binding assays have also been developed using mutant cell lines such as RMAS (in mice) and T2. These cells have low or no surface expression of MHC class I molecules; the $\alpha$ chains remain in the cytoplasm where they associate poorly with $\beta_2$-microglobulin. The defects in these lines can be restored by the addition of exogenous peptide. Binding of peptides within these cell lines stabilizes the MHC class I molecules which appear on the cell surface and thus enable CTL lysis. Consequently, any binding of peptides enhances surface expression of MHC I molecules at the cell surface which can be measured. It is assumed such peptides are potential CTL epitopes. Of the peptides identified by this method, there is a high probability that some will be natural CTL epitopes (see Section 6.4.4). T2 cells have also been transfected with other HLA molecules and the binding of peptides has been studied. A problem arises in differentiating between binding to the transfected HLA and any constitutive HLA A2.1 which is also present. Another novel way to analyze HLA-restricted responses relevant to humans is to immunize mice transgenic for a specific HLA haplotype and analyze the transgenic HLA-restricted responses.

### 6.4.3 Computer-aided identification of CTL epitopes

Considering the diversity of the MHC class I and class II molecules and the fact that different haplotypes may recognize different peptides, it is

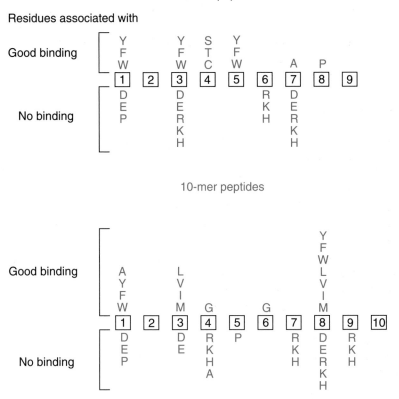

**Figure 6.17:** Residues strongly associated with good or poor HLA A2.1 binding. Based on the 160 synthesized 9-mer peptides of viral or tumor origin. Anchor residues in all peptides were L or M in position 2 and L, V or I in the C-terminal position 9 or 10. Reproduced from ref. [11] with permission from Cell Press.

surprising to find that algorithms can be developed which have any predictive value. Initially predictions were based on searching the primary amino acid sequence for (charged or Gly)–Φ–Φ–(polar or Gly) or (charged or Gly)–Φ–Φ–(Φ or Pro)-(polar or Gly) where Φ is a hydrophobic amino acid residue. Recently, a number of elegant X-ray crystallographic studies have detailed the structure of both human class I and class II molecules. In the case of HLA-A2.1 (class I), detailed studies have shown six different pockets in the peptide-binding groove. The two main pockets make contact with residues at position 2 and at the C terminus of the peptide. For A2.1 these 'anchor' residues are leucine at position 2 and valine or leucine at position 9 in the peptide. Based on these observations, *XLXXXXXX*V/L would be predicted to define a CTL epitope; however, this is only about 30% accurate and other residues, notably 1, 3 and 7, are important in A2.1 recognition of peptides. An extended motif taking into account secondary anchors claims an increased predictability to a level of 70% for HLA A2.1-binding epitopes.

It is now also possible to recover MHC class I molecules from cells, isolate the bound peptide and sequence it. This information coupled with *in vitro* binding assays will undoubtedly improve the predictability of binding epitopes still further. It should be noted that binding does not necessarily guarantee that an epitope will create a target for CTL because residues of the peptide also have to interact with the T-cell receptor on the CTL. A further theoretical possibility is that exogenous addition of a peptide may allow binding that would not normally have occurred naturally, because peptides are not necessarily processed and transported with equal efficiencies.

### 6.4.4 EBV LMP2

*Figure 6.18* shows 35 peptides predicted to be CTL epitopes for HLA-A2.1 from the EBV LMP2 gene. The algorithm used, $X[ILM]XXXXX[VL]$, where $X$ is any amino acid, is a slight modification of the one mentioned in Section 6.4.3. These 35 peptides were then tested in a T2-binding assay measuring relative binding efficiencies against the binding of a monoclonal W6-32 which binds to MHC class I framework. The specificities of CTL (both bulk cultures and T-cell clones) from donors that are latently infected with EBV and have an HLA A2.1 haplotype were determined by using the peptides to create targets. As can be seen, two CTL epitopes

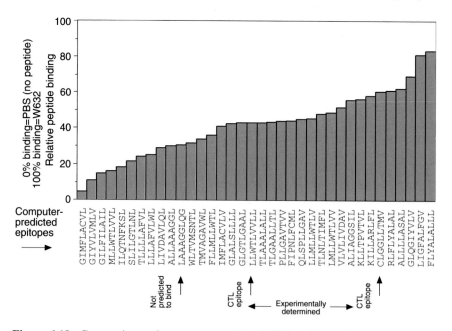

**Figure 6.18:** Comparison of computer-predicted CTL epitopes for EBV LMP2 and *in vitro* T2 binding assays with experimentally determined CTL epitopes. Unpublished data from Dr S.P. Lee, CRC Laboratories, Department of Cancer Studies, University of Birmingham, UK.

were determined experimentally, and one of these peptides was a relatively poor binder by T2 assay. Interestingly, several peptides bound more strongly to HLA A2.1 on the T2 cell line than did the CTL epitopes defined experimentally. It is still theoretically possible that some of these peptides may turn out to be CTL epitopes *in vivo* and, for technical reasons, *in vitro* stimulation required for generating CTL bulk cultures and clones did not have a proliferative effect on these specificities.

This example illustrates the limitations of current predictive and biochemical assay methods but, as techniques improve, the full power of this methodology will become apparent.

## References

1. Stone, R. (1993) *Science*, **262**, 833.
2. Hopp, T.P. and Woods, K.R. (1981) *Proc. Natl Acad. Sci. USA*, **78**, 3824.
3. Kyte, J. and Doolittle, R.F. (1982) *J. Mol. Biol.*, **157**, 105.
4. Chou, P.Y. and Fasman, G.D. (1978) in *Advances in Enzymology* (A. Meister, ed.). Wiley, New York, p. 45.
5. Garnier, J., Osguthorpe, D.J. and Robinson, B. (1979) *J. Mol. Biol.*, **120**, 97.
6. Jamieson, B.A. and Wolf, H. (1988) *CABIOS*, **4**, 181.
7. Nara, P.L. (1991) *FASEB J.*, **5**, 2443.
8. Tindle, R.W. *et al.* (1990) *J. Gen. Virol.*, **71**, 1347.
9. Wiley, D.C. and Skehel, J.J. (1990) in *Virology,* 2nd edn (B.N. Fields and D.M. Knipe, eds). Raven Press, New York.
10. Colman, P.M., Laver, W.G., Varghese, J.N., Baker, A.T., Tulloch, P.A., Air, G.M. and Webster, R.G. (1987) *Nature (Lond.)*, **326**, 358.
11. Rupert *et al.* (1993) *Cell*, **74**, 929.

# Chapter 7

# Developing new vaccines: inactivated and subunit approaches

## 7.1 Inactivated vaccines

Inactivated vaccines are made from virulent pathogens by destroying their infectivity while retaining their immunogenicity. Vaccines prepared in this way are relatively safe, and generally stimulate high levels of antibody against the pathogen's surface proteins but poor CTL responses. Often a primary course of vaccination comprises two or three injections, and further booster doses are required at a later date.

Care is taken to ensure that all infectious organisms are inactivated, and this is done with either β-propriolactone or formaldehyde. Whole inactivated pathogens or extracts of the pathogens can be used. Typical examples of inactivated vaccines are shown in *Table 7.1*.

An adaptation of this approach is the use of inactivated bacterial exotoxins such as those produced by tetanus and diphtheria. Protective antibodies react only with the toxin and not with the infectious bacterium. Such vaccines can be considered to be forerunners of subunit vaccines but they are more accurately represented by the capsular polysaccharide of *H. influenzae* B or *N. meningitidis*. In the past, manufacture of subunit vaccines has required biological procedures to grow the pathogen and physical or chemical techniques to isolate the required product. The current tendency is to use genetic engineering techniques.

## 7.2 Expression systems and subunit vaccines

The production of a subunit vaccine for HBV by expression of the HBsAg in both yeast and eukaryotic cell lines (see Chapter 10) is one of the most

**Table 7.1:** Inactivated vaccines available in the UK

| Vaccine | Inactivation | High risk to UK residents | Administration |
|---|---|---|---|
| Polio | Formaldehyde | Birth onward | At 2, 3, 4 months old |
| Pertussis | Formaldehyde | Children < 6 months | At 2, 3, 4 months old |
| Rabies | β-propriolactone | Individuals in contact with imported animals or post-exposure if bitten by wild animals | 0, 3, 7, 14, 30 and 90 days |
| Japanese encephalitis virus (JEV) | Formaldehyde | Travelers to areas endemic for JEV (highest risk after monsoon) | 0, 7–14 and 28 days |
| Cholera | Heat killed, phenol preserved | Travelers to endemic areas | 0 and 28 days |
| Influenza | Formaldehyde, solvent or detergent extract | Elderly, chronic respiratory disease including asthmatics, other medical conditions | A single injection, if primary immunization 0, 28–36 days later |

frequently cited successes of recombinant DNA technology. Improvements in the technology for expressing and purifying proteins, together with the fact that they tend to have very low complication rates, have persuaded manufacturers to favor the subunit approach over other strategies for vaccine production.

### 7.2.1 Proteins expressed by prokaryotes

Advantages of expressing genes in *E. coli* include: (1) a wealth of information available about gene expression and regulation in the organism; (2) the ease of manipulation and speed of growth; (3) well-established technology for large-scale production. There are two main disadvantages associated with intracellular over-expression of proteins in *E. coli*. The first is the tendency of foreign proteins to form insoluble inclusion bodies which comprise masses of associated, unfolded polypeptide chains. In some cases this material can be refolded but it is often difficult, time-consuming and often leads to decreased yields of product. Modification of growth conditions can improve solubility, and protein secretion systems are now proving to be effective, but the production of soluble active protein remains the greatest problem with expression in *E. coli*.

A second disadvantage for *E. coli* as a host organism is its inability to carry out post-translational modifications that are naturally found on many proteins (*Table 7.2*). A further problem for proteins required for pharmaceutical use is the need to remove LPS and pyrogens (endotoxins) produced by bacteria.

**Table 7.2:** Post-translational modification in various expression systems

| Modification[a] | Expression system | | | | |
|---|---|---|---|---|---|
| | E. coli | Yeast | Mammalian cells | Baculovirus (insect cells) | Transgenic animals |
| Glycosylation | No | Yes[c] | Yes | Yes[d] | Yes |
| Myristylation | No[b] | Yes | Yes | Yes | Yes |
| Phosphorylation | No | Yes | Yes | Yes | Yes |
| Heterodimer assembly | Yes | Yes | Yes | Yes | Yes |

[a]Other post-translational modifications could be included in this table, e.g. palmitoylation, acylation, methionine removal, signal cleavage.
[b]When the yeast enzyme *N*-myristyl transferase is expressed in *E. coli* it will myristylate foreign genes.
[c]*S. cereviseae* hyperglycosylates some glycoproteins.
[d]Insect cell glycosylation differs from mammalian cell glycosylation. Complex sugars are added only inefficiently to the core glycosylated protein.

To obtain good yields of product it is necessary to use a system which allows control of expression, especially if the product is in any way toxic to the bacterial cell. Several well-characterized strong promoters which can be regulated to give expression when desired are widely available (*Table 7.3*). Even if strong promoters are used and regulated expression achieved, there is no guarantee that the proteins will be produced at even reasonable levels as active, soluble or antigenic proteins. For example, some proteins such as human TNF-$\alpha$ are soluble under all conditions, some like IFN-$\gamma$ are partially soluble, whereas some, including human tissue inhibitor metalloproteases (TIMP), are totally insoluble. Virus glycoproteins, including influenza HA and HBsAg, have proved to be particularly difficult to express at high levels in *E. coli*. However HBV core antigen (HBcAg) can be expressed at very high levels, approaching 20% of total cell protein. This unpredictability means that it is necessary to establish experimentally the optimal conditions for expression of each specific gene. Part of the difficulty of expressing virus glycoproteins appears to be related to hydrophobic regions of the molecules such as the transmembrane anchor region. Removal of the anchor region of HSV glycoprotein gB significantly improved the level of expression of the protein.

**Table 7.3:** Promoters used for regulated expression in *E. coli*

| Promoter | Repressor | Inducer |
|---|---|---|
| *trp* | *trp* | IAA (3-$\beta$-indoylacrylic acid) |
| *lac/trp* | *lacI* | IPTG |
| P$_L$ | c$^{857}$I | Temperature shift |
| *RecA* | *lexA* | Nalidixic acid |
| T7 | *lacI* | IPTG[a] |

[a]The T7 RNA polymerase gene is expressed from a *lac* promoter and hence can be induced with IPTG.

Some of the solubility problems can be overcome by expressing the target protein as a fusion with a stable more soluble bacterial protein. Such fusions are valuable research tools but are unlikely to be used for large-scale production of proteins. The target protein may be isolated by incorporation of a protease site between the target protein and the fusion partner. Protein tags can also be added to the ends of molecules for purification with monoclonal antibodies or simply to allow detection of proteins (*Table 7.4*).

A variety of other sequence elements and diverse factors have been used to obtain the desired degree of control, expression and solubility. These include modified ribosome binding sites, transcription termination sequences, the choice of vector with optimal copy number, growth at different temperatures and dual origin vectors for copy-number amplification.

Despite the difficulties of expressing many viral glycoproteins with potential as vaccine antigens, the simplicity of manipulation and scale-up make *E. coli* very attractive as an expression system. It is clear that antigens for experimental use and diagnosis will continue to be made in bacteria, and a number of engineered bacteria expressing antigens may be used as live, attenuated vaccines (see Sections 8.2, 8.7 and 8.8). However, it is likely that other systems more capable of authentic post-translational modifications will be used for the production of the majority of subunit vaccines.

**Table 7.4:** Fusion protein and tag systems used in *E. coli*

| Fusion/tag | Purification |
| --- | --- |
| His ( × 6) | Metal-chelating columns |
| 'Flag'[a] | Monoclonal antibody |
| HSV gD peptide | Monoclonal antibody |
| *E. coli* protein that is naturally biotinylated | Avidin–agarose |
| β-galactosidase | APTG–Sepharose |
| Maltose-binding protein | Amylose agarose bead |
| Glutathione-*S*-transferase of *Schistosoma japonicum* | Glutathione agarose |

[a]Commercial name for a peptide sequence.

### 7.2.2 Proteins expressed by yeast

Several key features have contributed to the continued choice of yeast as an expression system: (1) the use of yeasts in the food industry has meant that there are healthy laboratory and industrial strains of yeast with low levels of endotoxin present in purified products; there is a high yield fermentation technology developed with associated hardware providing a simple scale-up from bench level to fermentation scale; (2) expression systems are well worked out with appropriate selectable markers, strong regulatable promoters and the choice of integrating into the yeast genome or replicating autonomously as high copy-number plasmids; (3) many of the post-translational modifications carried out in mammalian cell culture are also found in yeast, including particle assembly and efficient secretion (see *Table 7.2*).

These advantages were further emphasized when it was found that HBsAg assembled spontaneously into characteristic antigenic 20–22 nm particles which could be expressed at commercially acceptable levels in *S. cerevisiae*. This led to the eventual use of yeast-derived HBsAg as the first (and so far only) widely licensed recombinant human vaccine (*Table 7.5*). In a similar flagship development, the secretion-efficient yeast *Kluyveromyces lactis* was used in the food industry to express the milk-clotting enzyme, chymosin.

For high-level, stable expression of foreign genes in yeast, plasmids which replicate extrachromosomally are used rather than integrating vectors (see *Figure 7.1*). YEp24 contains the *URA3* marker gene for selection of transformed yeast cells with a mutation in the *URA3* gene using media deficient in uracil. Other selectable markers include *LEU2*, *HIS4*, *TRP1* and *LYS2* for propagation in leucine-, hisitidine-, tryptophan- and lysine-deficient media, respectively. The vector contains a pBR322 backbone so that recombinant DNA manipulation can be done in *E. coli*. The 2 μ element allows stable high copy number replication of plasmid during growth of *S. cerevisiae* and is also responsible for faithful partitioning of plasmids at cell division. Plasmids such as YIp5 contain a selectable marker and integrate into the yeast genome at the marker locus. YEp24 and YIp5 do not contain yeast promoters but would form the basis of a vector such as pHBS-16.

A variety of strong yeast promoters including those from the alcohol dehydrogenase I and II, enolase, glyceraldehyde-3-phosphate dehydrogenase, phosphoglycerate kinase and acid phosphatase genes have been used to drive foreign gene expression. As yeast strains in current use have been derived over a long period of time for their ability to carry out

**Table 7.5:** Examples of antigens from pathogens expressed in yeast

| Product | Comments |
| --- | --- |
| HBsAg | Licensed vaccine, assembles into particles |
| HBsAg–epitope fusions (HSV gD or malarial CSP) | Particles formed with epitopes of the fusion partner exposed at the surface |
| Malaria *Plasmodium vivax* CSP | – |
| Malaria *P. falciparum* SERA proteins | Immunogenicity studied in rodents |
| HCV human superoxide dismutase fusion proteins | Used in HCV diagnosis |
| HIV-1 gp120 | Chiron–Biocine product in phase I/II trials in USA |
| SIV *gag/env*–TY particle fusions | Stimulates IgA and mucosal protection of experimental animals |
| HIV-1 *env*– and *gag*–TY particle fusions | TY–*gag* fusion vaccine in trials in UK. Assembles into highly antigenic particles |

CSP, circumsporozoite protein; HCV, hepatitis C virus; SIV, simian immunodeficiency virus; SERA, serum erythrocyte ring antigen.

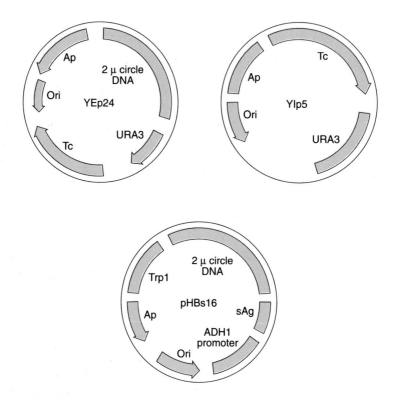

**Figure 7.1:** Plasmid vectors for use in yeast. *URA3* and *TRP1* allow selection in yeast. ADH1, alcohol dehydrogenase gene; Ap, ampicillin; ori, bacterial origin of replication; sAg, hepatitis B virus surface antigen; Tc, tetracycline resistance gene.

glycolysis, it is not surprising that the strong promoters initially identified were also promoters for those genes involved in glycolysis.

A further sophistication incorporated into yeast expression systems is the ability to secrete the recombinant product. Although some higher eukaryotic signal sequences can function in yeast, others give heterogeneous products. Consequently, the standard procedure is to fuse the foreign gene of interest with a secretion signal in yeast such as the α-mating factor signal/leader sequence. The fusion protein is expressed and directed to the secretory pathway where the signal sequence is lost and the foreign gene is secreted (see *Table 7.2*).

Retrotransposons (TY elements) of *S. cerevisiae* are a family of dispersed repetitive DNA sequence elements. They share functional and structural similarities with retrovirus proviral genomes and produce virus-like particles (VLP) as their primary translation product. It has been found that even quite large foreign gene products fused to the VLP do not disrupt their structure. Hybrid HIV *gag* and *env* particles are readily purified and are immunogenic in laboratory animals. Hybrid TY-SIV *gag* p27 VLP covalently linked to cholera toxin B subunit will also stimulate mucosal immunity in a primate model.

Virus-specific glycoproteins are one of the prime targets for neutralizing antibodies to enveloped viruses. Yeast hyperglycosylates many heavily glycosylated virus glycoproteins such as EBV gp340 and HIV gp120. In the case of HIV gp120, this appears not to matter as high-titer antibodies can be generated against the form expressed in yeast, but gp340 expressed in both *S. cerevisiae* and *Pichia pastoris* is only poorly antigenic. With this limitation in mind, virus glycoproteins are more often expressed in mammalian cell culture. The advantages of expressing vaccine antigens in yeast undoubtedly outweigh the limitations, and it is likely that more subunit vaccines will be derived by expression of genes in yeast.

### 7.2.3 Proteins expressed in mammalian cell culture

As with other vector systems an increasing understanding of the transcriptional and translational control of the host has led to efficient systems for the high-level production of proteins. Other requirements such as strong promoters, selection systems and means for improving expression, such as gene amplification, have all been identified and are illustrated in *Table 7.6*.

Two systems have been widely used for producing proteins in mammalian cells based either on COS cells or on Chinese hamster ovary (CHO) cell lines. Many CHO cell mutants are available and have been used to make selection systems which allow foreign genes to be transferred and selected COS cells are not appropriate for large-scale production of protein and are used primarily for studies of wild-type and mutant proteins. (For further reading on selection and expression systems, see ref. [1]). Many candidate vaccines have now been prepared using a CHO cell line deficient in the gene for dihydrofolate reductase (*dhfr*). The gene to be expressed under appropriate promoter control is co-transformed with a plasmid construct that contains the *dhfr* gene. DHFR detoxifies the drug methotrexate (MTX) by cleaving MTX into biologically inactive subunits. Increasing concentrations of MTX amplify the *dhfr* gene along with the gene of interest and, as a consequence, amplify the levels of expression of the foreign gene.

Several virus vector sytems have attracted attention mainly because they maintain their genomes episomally. Bovine papillomavirus (BPV) vectors in particular have been used to express commercially viable levels of biologically active proteins. Cloned BPV DNA will transform mouse NIH 3T3 or C-127 cells. Visible transformed foci (*Figure 7.2b*) on a background of normal cells can be isolated and grown as continuous cell lines. Plasmids containing BPV and a gene to be expressed under control of a eukaryotic promoter (*Figure 7.2a*) are transfected into C-127 cells, and transformed foci picked, cloned and expanded into cell lines. Using this methodology, candidate vaccines for HBV (see Chapter 10) and EBV have been produced (see Chapter 9).

**Table 7.6:** Expression levels and utilities of various mammalian gene transfer and expression systems

| Cell line | Mode of DNA transfer | Optimal expression level ($\mu$g ml$^{-1}$) | Primary utility |
|---|---|---|---|
| *Human/primate cells* | | | |
| CV-1 | SV40 infection | 1–10 | Expression of wild-type and mutant proteins |
| CV-1/293 | Adenovirus infection | 1–10 | |
| COS | Transient lipo/DEAE DNA transfection | 1 | Cloning by expression in mammalian cells; rapid characterization of cDNA clones; expression of mutant proteins |
| CV-1 | Transient DNA transfection | 0.05 | |
| *Rodent cells* | | | |
| C-127 | BPV stable transformant | 1–5 | High-level constitutive expression |
| 3T3 | Retrovirus infection | 0.1–0.5 | Gene transfer into animals; expression in different cell types (with amphotropic packaging line) |
| CHO-DHFR | Stable DHFR + transformant | 0.01–0.05 | High-level constitutive expression |
| | Amplified with MTX | 10 | |
| Various | Vaccinia vector | 1 | Vaccines? CTL targets; can be used to isolate proteins |
| | EBV vector | NA | Cloning by expression; expression of wild-type and mutant proteins |

It is difficult to compare these systems directly because different genes or promoters or constructs are often used to analyze proteins produced in the various systems. Most cell lines used with BPV vectors require attachment, and as yet there are no suspension variants of C-127 cells which would decrease production and purification costs. The HBsAg with its own promoter in COS and CHO cells is synthesized at between $1 \times 10^8$ and $3.3 \times 10^8$ molecules per cell per day compared with $1.6 \times 10^6$ molecules per cell per day using BPV expression systems. However, the difference between the two systems is less marked for other promoters and genes. CHO cells amplified with DHFR selection and MTX can lose the co-expressed foreign gene on prolonged passage. CHO cells also tend to give different glycosylation patterns on some gene products with higher levels of sialic acid than the authentic molecules. It is, however, probably not the scientific merits or possibly even the financial considerations that will determine whether CHO or C-127 cells are more appropriate for the production of

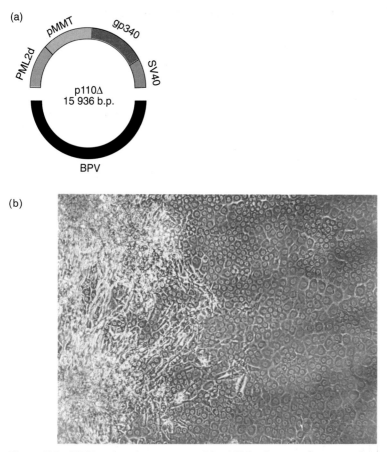

**Figure 7.2:** BPV expression vectors. (a) p110Δ when used to transform mouse C-127 cells, will replicate episomally due to the BPV sequences present and will express in the cell supernatant a truncated version of the EBV glycoprotein gp340. pMMT, mouse metallothionein protein promoter element; SV40, poly A sequence. (b) A focus of transformed cells which contain the BPV construct can be seen on a background of normal cells.

biological products. It is the regulatory authorities who to some extent determine the priorities. Unlike BPV-transformed C-127s, CHO cells do not form tumors in nude mice and do not contain virus genes capable of transformation. As a result, products derived from CHO cell lines are being widely licensed in the USA and Europe whereas those from BPV-transformed lines seem acceptable to European regulatory authorities (*Table 7.7*).

### 7.2.4 Baculovirus vectors

Methods for insertion and expression of genes in baculoviruses under the control of strong virus-specific promoters have been described which are akin to the methodology associated with the construction of recombinant

**Table 7.7:** Selected examples of vaccines expressed in mammalian cell culture

| Gene expressed | System used | Comment |
|---|---|---|
| HBsAg | C-127/BPV | Phase I/II clinical trials |
| EBV gp340 | C-127/BPV | Preproduction development stage |
| HSV gD and gB | CHO/DHFR amplified | Phase II clinical trials |
| HBsAg-PreS1 | CHO/DHFR amplified | Fully licensed vaccine |
| HIV gp160 | CHO/DHFR amplified | Phase I/II clinical trials |

vaccinia viruses (see Chapter 8). The initial attraction of baculoviruses was reports of very high levels of gene expression in insect cells – as much as 30% of total cell protein. This was achieved by using the polyhedron promoter of *Autoradiographica californica* multiple capsid nuclear polyhedrosis virus. Since then several other promoters (for example, p10 and basic protein promoter) have been described which also give high levels of foreign gene product. Although many proteins are not made at this level, the system still makes large amounts of protein which have proved in almost all cases to be biologically active.

As indicated in *Table 7.8*, many eukaryotic post-translational modifications are carried out in insect cells, the only apparent short-coming of the system being that the cells add less carbohydrate to core glycosylated proteins. The biological significance of this is unclear, and, where expressed proteins have been used as immunogens, good protective responses have been elicited.

### 7.2.5 Further developments

A number of interesting concepts for the expression of foreign gene products have been advanced in the last few years. Transgenic domestic animals are now being proposed as an alternative way of producing

**Table 7.8:** Baculovirus-expressed immunogens and particles

| Virus | Antigen |
|---|---|
| Human parainfluenza virus type 3 | F glycoprotein, HA/NA protein |
| Human RSV | F glycoprotein, G |
| Rift valley fever virus | G1, G2 |
| HSV | gD |
| Rabies virus | G |
| HIV | gp120/160 (MicroGenSys material phase I/II trial in USA) |
| | Empty virus particles formed from *gag* coding region; co-infection with baculovirus expressing gp160 led to the production of particles with gp160-containing envelope |
| Poliovirus | Capsids produced from cDNA clone lacking 5'-untranslated region |
| HBV | Co-expression of HBsAg and HBcAg[a] |

[a]For details see ref. [2].

recombinant proteins. Work is in progress to achieve high-level expression; α1 anti-trypsin was produced in the milk of transgenic mice using the β-lactoglobulin promoter at levels of 0.4–12.45 mg $l^{-1}$. When the same construct was transferred to transgenic sheep, levels of 35 g $l^{-1}$ were achieved (about 50% of total protein). Since this yield can be sustained throughout lactation, the yield of α1 anti-trypsin could exceed 10 kg per animal. One small herd of lactating sheep expressing human clotting factor XI at this level would supply the world's total requirement for factor XI. Interestingly, the α1 anti-trypsin was fully N-glycosylated with biological activity indistinguishable from human plasma-derived α1 anti-trypsin, even though normally less than 10% of endogenous milk protein is glycosylated, and this represents O-linked glycosylation of κ-casein. It is only a short jump to propose that vaccine antigens could be produced by this technology. It will be interesting to see if the promise of the technology is fulfilled and if vaccine antigens are expressed in this way.

Another exciting prospect is the engineering of plants to express molecular determinants which could stimulate immune responses. Cholera toxin B-chain and chimeric proteins derived from the B chain are being expressed in plants to evaluate anti-toxin vaccines delivered to the gut mucosal immune system, and it will be particularly interesting to see if mucosal immunity can be stimulated. The implications for developing countries are particularly striking. Incorporation of vaccine antigens into food crops would ease the burden of the cost of vaccination and simplify the administration of vaccine greatly. An enormous amount of work is needed to see if this approach is possible but if it were to be sucessful it could revolutionize vaccine practice. Our morning bowl of muesli might in the future be both nutritious and give us a booster dose of vaccine!

## 7.3 Peptide vaccines

This approach is the ultimate in the reductionist approach to vaccines. As mentioned in Chapter 6, it is possible to identify the epitopes within a protein that can induce neutralizing antibody or epitopes that are important in T-cell responses to antigens. Numerous studies have shown that linear synthetic peptides corresponding to part of a protein can induce antibodies that bind the intact protein. Antipeptide sera can be generated that react:

(1) only with the immunizing peptide;
(2) with the peptide and native protein but do not compete with antibodies raised against the native protein (peptide-specific); or
(3) with peptide, native protein and compete with antibodies raised to the native protein (sequential determinant-specific).

In terms of vaccines those peptides that induce neutralizing antibodies are of most interest (some antipeptide sera that are capable of mediating ADCC may not be virus-neutralizing *in vitro*). Thus peptides that elicit peptide-specific or determinant-specific antisera are more promising because they react with native protein.

Chemical synthesis of peptides is now straightforward, and with appropriate adjuvant or conjugation to carrier proteins they can induce immune responses that are sufficient to give protection against the organism from which the peptide was derived. For example a 27 amino acid synthetic peptide from HBsAg (amino acids 284–311) coupled to keyhole limpet hemocyanin (KLH) will protect chimpanzees against HBV challenge. Interestingly sera to this peptide will bind surface antigen and peptide but will not compete with anti-native monoclonal antibodies.

The original concept of a peptide vaccine was a B-cell epitope conjugated to a heterologous protein carrier such as tetanus toxoid. Such a simplistic notion has numerous limitations which have subsequently become apparent. They include carrier-induced suppression, the lack of well-defined carrier proteins suitable for human use, and the realization that priming of the T-cell memory response relevant to the pathogen is not merely beneficial but essential.

Only relatively recently has the importance of including antigen-relevant T-cell recognition sites been fully appreciated. The challenge to the designer of synthetic peptide vaccines which elicit B-cell recognition is to overcome the conformational nature of many B-cell epitopes either by selecting nonconformational epitopes or by learning how to give structure and mimic conformational epitopes. It is imperative that a synthetic T-cell epitope is capable of priming a memory T-cell response that can be recalled by a determinant of the pathogen. The challenge in terms of T-cell recognition is to provide a sufficient diversity of recognition sites to immunize outbred populations with many HLA haplotypes.

It is possible to some extent to meet these challenges and mimic both B- and T-cell responses to antigens with peptides but the rules of peptide design are not yet fully clear. *Table 7.9* shows a number of composite synthetic peptide immunogens and the response induced in mice. It is clear from this that changes in the order of T- and B-cell epitopes as well as the

**Table 7.9:** Composite synthetic immunogens for HBV containing different B- and T-cell epitopes

| T site | Antipeptide | | | |
| --- | --- | --- | --- | --- |
| | B site (preS2) | T site | B site | Anti-native |
| | 133–145 | 0 | 0 | 0 |
| PreS2 120–132–133–145 | | 0 | + + | + + |
| PreS1 12–21–133–151 | | 0 | + + | + + |
| HBc 120–140–133–143 | | + + + | + + | + + |
| PreS2 151–174–133–143 (amino terminus) | | + + + | + + | +, – |
| PreS2 151–174–133–143 (carboxyl terminus) | | 0 | + + + + | + + + |

use of different T-cell epitopes alter the fine specificity of the immune response, and it will be difficult to guarantee how a given peptide with T- and B-cell epitopes will behave in a diverse recipient population.

Although a 'universal' T-cell epitope probably does not exist, the 148–174 sequence of the preS2 region represents a relatively short sequence (27 amino acids) that is recognized by T cells in the context of all seven MHC haplotypes tested. Rather than this sequence representing one T-cell site recognized by all haplotypes this sequence contains numerous sites recognized uniquely by each MHC type. Presumably other pathogens have similar sequences which can be included in synthetic vaccines to serve as functional T-cell epitopes in an outbred population.

Modifications to the basic technique of coupling a peptide to a protein and using it as a vaccine are all designed to improve the immune response and include:

(1) incorporation of a helper T-cell epitope within the peptide and use of the peptide on its own;
(2) cyclization of the peptide to improve antigenicity;
(3) incorporation of the peptide into antigenic regions of other proteins, (such as the poliovirus capsid (see Section 8.9), bacterial fimbriae (see Section 8.8) or HBcAg illustrated in *Table 7.10*. The advantage of incorporation into HBcAg or poliovirus capsids rather than other proteins is that they both form particles which by their very nature are highly immunogenic); and
(4) construction of higher-order peptide structures with multiple copies of the peptide. This has been achieved in a number of ways; the most promising are polymerization and branching (*Figure 7.3*).

A good example of the polymerization approach comes from attempts to develop a vaccine for malaria. Dr M. Patarroyo designed a molecule which has sequences from three different asexual stage proteins (85, 55 and

**Table 7.10:** Incorporation of peptides into hepatitis B virus core antigen (HBcAg)

| Sequence inserted into HBcAg | Site of insertion |
|---|---|
| HBsAg S285–330 | COOH terminus |
| HBsAg S285–339 | COOH terminus |
| HBsAg preS1 1–20 | COOH terminus |
| HBsAg preS1 1–36 | COOH terminus |
| HBsAg preS2 120–145 | COOH terminus |
| HBsAg preS1 27–53 | Internal |
| HBsAg preS1 12–47 | $NH_2$ terminus |
| FMDV VP1 142–160 | COOH terminus |
| FMDV VP1 142–160 | Internal |
| FMDV VP1 142–160 | $NH_2$ terminus |

FMDV = foot and mouth disease virus; VP1 is the largest of the virus capsid proteins. The table is not an exhaustive list of all examples of epitopes inserted into HBcAg but illustrates the potential power of the system

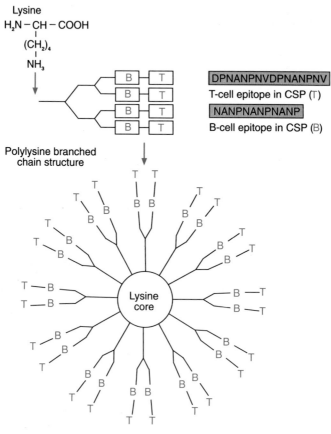

Multiple antigen peptide

**Figure 7.3:** Example of the construction of a branched-chain multiple antigenic peptide structure. In this case the T- and B-cell epitopes from a malarial circumsporozoite surface antigen (CSP) are linked to lysine to form a highly polymeric structure with a lysine core and many copies of the CSP epitopes.

35 kDa) linked by pNANP sequences present on the sporozoite stage of the malarial parasite. The basic monomer peptide is 45 amino acids in length. It is allowed to polymerize under controlled conditions, linking monomers end-to-end through disulfide bonds and side-to-side through sulfoxyl bonds. The polymer Spf66 is adsorbed on to aluminum hydroxide and used as a vaccine. Although there has been some controversy surrounding the reproducibility of results and the design of trials, a recent phase III trial [4] does seem to show some protective effect of the vaccination.

*Table 7.11* shows some of the advantages of peptide vaccines. These advantages stem from peptides being a precisely defined product which are simple and cheap to produce. As vaccines they elicit a precise immune response and can be used safely.

**Table 7.11:** Advantages of peptide vaccines

| Advantages |
| --- |
| 1. Chemically defined and unlimited source of material |
| 2. Stable indefinitely, reducing problems of delivery and storage |
| 3. No infectious agent present |
| 4. No large-scale production plant required |
| 5. No downstream processing required, consequently inexpensive to produce |
| 6. Can be designed to stimulate specific B and T cell-mediated responses |

The major disadvantage often cited for peptide vaccines is based on the fact that usually only one peptide is used. Many pathogens exhibit extreme variation in the antigenic proteins of the agent, for example HIV-1 gp160. Could a single epitope, or even multiple epitopes, be found that protect in all cases in the face of the extreme antigenic variation and fulfil the requirement to stimulate T-cell memory in an outbred population? There is also the possibility that a single epitope peptide may elicit monospecific antipeptide antibodies from which natural variants may escape by point mutation. In particular, the inherent mutation rates of many RNA viruses are so high that they readily generate neutralization-escape mutants. Despite these reservations, the ease of production, stability and safety of peptide vaccines ensure that they will receive much attention in the future.

## 7.4 Anti-idiotypes

It has been suggested that anti-idiotype (anti-Id) antibodies may make effective vaccines. This is based on the finding that antibodies themselves can act as immunogens. An immune response raised against the unique antigen-combining site of an antibody is termed an anti-idiotypic response. This second antibody may bear a structural resemblance to the original antigen. When this occurs the anti-idiotypic antibody (monoclonal or polyclonal) may be able to induce an antibody response that recognizes the original antigen and hence to act as a vaccine (see *Figure 7.4*) Such protection has been demonstrated in a variety of animal model systems; probably the best demonstration of the potential of this approach is the protection of chimpanzees from HBV-associated diseases by previous immunization with anti-Id.

There are several circumstances where anti-Id immunization offers an advantage over traditional approaches. These are shared with other subunit vaccine strategies, particularly peptide vaccines. They are:

(1) if the antigen is difficult to obtain (i.e. when the infectious agent is hazardous or cannot be grown *in vitro*);

(2) if the attenuated vaccines have high reversion frequencies or possess genes that may be involved in oncogenesis (any problems vaccinating immunocompromised individuals with live vaccines would also be avoided);

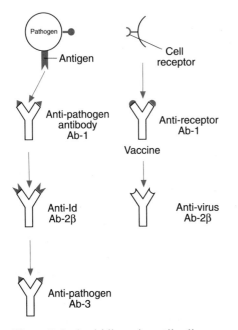

**Figure 7.4:** Anti-idiotypic antibodies as vaccines. Two approaches for generating Id-based vaccine strategies are depicted. The more classical approach involves an Ab-2β preparation that mimics a neutralizing epitope or κ pathogen (■) (may or may not involve the receptor-binding site). The Ab-2β is the vaccine and induces a response, Ab3, that mimics Ab1 and binds the original antigen. In the second approach the Id on the anti-receptor antibody is the vaccine and induces an Ab2 response that serologically mimics the receptor molecule and binds the virus at its site of action.

(3) if a single epitope can confer protection but other epitopes of the whole molecule might induce autoimmunity;

(4) if organisms display wide genetic diversity but a single cellular receptor (an anti-Id response could theoretically induce a serological response that mimics the receptor and binds the infectious agent at its receptor-binding site); and

(5) the ability of anti-Id to mimic nonproteinaceous epitopes such as carbohydrate, lipid or glycolipid (none of which can yet be produced easily) to act as subunit vaccines.

There are a number of disadvantages to the anti-Id vaccine strategy, in particular the restriction of the vaccine to a single epitope, or a few epitopes administered together, which may not be enough to protect against some organisms. Another significant limitation associated with its multiple use is the possibility that antibodies to C region immunoglobin determinants will arise. This might then prejudice subsequent immunizations, and antibody–antigen complex formation could lead to immunopathological damage.

Despite these limitations, and although this approach offers little advantage over other recombinant sources of antigen other than for nonproteinaceous epitopes, it may prove to be an important adjunct to other strategies for immunization.

## 7.5 Genetic immunization

It has been found recently that intramuscular (i.m.) injection of nonreplicating plasmid DNA encoding influenza A virus genes can generate both humoral and cell-mediated protective immunity against influenza in mice and ferrets. In mice, immunization with DNA (or mRNA) encoding group-specific nucleoprotein induced CTL and protective immunity to different virus strains. Such CTL responses have been observed to last for at least 6 months. After immunization of nonhuman primates with the influenza HA gene, production of antibodies against the HA was maintained for a similar period. Thus this technology has the potential to generate both long-lived strain-specific antibodies and broadly cross-reactive cell-mediated immune responses capable of conferring protection against diverse virus strains. Recent studies have also described the use of this technology to induce anti-HIV immune responses including B-cell and T-cell immune responses in several animal species.

Other methods for introducing the DNA into cells that have proved successful include:

(1) a 'gene gun' for projecting DNA-covered gold particles directly into cells of living animals (the DNA presumably dissociates and can then be transcribed);
(2) i.m. injection with a pneumatic gun;
(3) complexes with liposomes administered either intravenously (i.v.) or as an aerosol; and
(4) a complex with asialo-orosomucoid poly-L-lysine (i.v. administration targets the complex to asialoglycoprotein receptors, particularly those of the hepatocyte).

A number of aspects of these observations have yet to be investigated. The introduced DNA has been shown to persist episomally for up to 18 months in mouse muscle, but how is this achieved? Myoblasts are particularly quiescent, and as such it is possible that the DNA is not lost or integrated because cell division does not occur. Which cells are presenting the antigen? If the antigen is expressed and an immune response is generated why are the cells expressing the antigen not recognized and destroyed by the induced immune response? Although it has been the subject of a fair amount of current research, the general applicability of the approach has yet to be established. It is unlikely that DNA vaccines will be used in the next 10 years although some of the associated technology may be useful in gene therapy.

## 7.6 Issues of presentation

### 7.6.1 Adjuvants

Purified, recombinant subunit and synthetic antigens are often poorly immunogenic on their own and require the simultaneous addition of other substances to induce effective immunity. A substance that can enhance the immunogenicity of an antigen is considered an adjuvant, and literally hundreds of such substances have been described. Disappointingly, the few adjuvants acceptable for use in man often do not sufficiently improve immune responses to give lasting protection. As a consequence the design and testing of systems to improve immune responses is an area of intense activity.

The only artificial adjuvants widely licensed are aluminum salts, often referred to as 'alum'. Aluminum salts are very safe but they do have a number of disadvantages. These include:

(1) variations in potency between different batches due to uncontrollable reactions between antigen and the aluminum salts;
(2) the requirement for refrigeration (complicating delivery in developing countries) because they cannot be frozen or lyophilized;
(3) they occasionally produce abscesses or nodules;
(4) failure to work with certain antigens; and
(5) at best their immunostimulating properties are limited, particularly with respect to the production of CMI.

A variety of strategies have been employed to enhance the immunogenicity of antigens, often two or more materials from different sources being combined in the hope of gaining additive or synergistic responses. Edelman and Tacket have approached this question in a useful way by classifying immunostimulatory materials into three categories: adjuvants, carriers for antigens, and vehicles. Examples of adjuvants include aluminum salts, saponin, muramyl di- and tri-peptides, monophosphoryl lipid A, *B. pertussis* and various cytokines including IL-12. Carriers include bacterial toxoids such as inactivated tetanus and cholera toxins, fatty acids, live vectors such as polio chimeras and hybrid proteins that form particulates, for example yeast retrotransposon hybrid TY particles and HBcAg particles. Vehicles, one or more of which is a component of most of the modern synthetic vaccines, consist of mineral oil emulsions (Freund's), vegetable oil emulsions (peanut oil), nonionic block co-polymer surfactants, squalene or squalane, liposomes and biodegradable polymer microspheres. Vehicles by themselves have independent immunostimulatory properties and mixtures of vehicles with adjuvants are referred to as adjuvant formulations.

*Table 7.12* shows details of a trial set up in early 1993 to test the relative efficacy of a number of different adjuvant formulations with one antigen (gp120). These types of data are in short supply and even in this trial several of the most promising adjuvant formulations have not been tested. Of particular interest is a saponin derivative, QS21, developed by Cambridge Biotech, which is soluble and stable to both freezing and thawing. This adjuvant has been used successfully in a subunit vaccine for feline leukemia virus. A Ribi Corporation formulation of monophosphoryl lipid A with mycobacteria cell walls has been tested clinically with a candidate malaria vaccine and also looks promising.

It should be noted that regulatory authorities have not licensed any adjuvant 'on its own', but they have all been part of a vaccine formulation. However, the widespread use of alum has placed it in a special position from the regulatory standpoint, all other adjuvants being considered experimental. To date, the USA FDA has never fully licensed a vaccine containing any experimental adjuvant.

### 7.6.2 Carriers

As mentioned previously, carriers are a diverse group of agents ranging from tetanus toxoid conjugated with capsular polysaccharide to recombinant polioviruses. They include TY particles (see Section 7.2.2), HBcAg particles (see *Table 7.10*), empty virus particles (see *Table 7.8*), liposomes, immunostimulatory complexes (ISCOMs), microcapsules and protein 'cochleates'. Immune responses to virus particles tend to be vigorous compared to the same amount of isolated antigen. A deliberate design feature of many of these carriers is to present numerous copies of an antigen or epitope in a particulate structure and, as a consequence, in some way to mimic virus particles.

**Table 7.12:** HIV vaccine and adjuvant formulation trial

| Adjuvant | Adjuvant developer |
| --- | --- |
| Alum (Alhydrogel) | |
| Water-in-squalene-in-water dispersion stabilized with polysorbate 80 (Tween 80) and containing monophosphoryl lipid A | Ribi Immunochem |
| Liposomes containing dimyristoyl phophatidylcholine, dimyristoyl phophatidylglycerol, cholesterol and monophosphoryl lipid A, all absorbed on to alum | Walter Reed Army Institute of Research |
| Squalene-in-water microfluidized emulsion stabilized with polysorbate 80 and sorbitan trioleate (MF59) | Chiron/Biocine |
| MF59 emulsion plus muramyltripeptide-phosphatidylethanolamine | Chiron/Biocine |
| Squalene-in-water microfluidized emulsion stabilized with polysorbate 80 and containing a nonionic block co-polymer L-121 (SAF-M) | Syntex, Chiron/Biocine |
| SAF-M emulsion also containing threonyl muramyl dipeptide | Syntex, Chiron/Biocine |

Many types of liposome have been tested over the past 20 years, with promising results. Recently, spherical, unilamellar vesicles, termed virosomes, containing HA glycoprotein of influenza A and a hepatitis A virus vaccine have been tested in humans. High titers of antibody were obtained after a single dose of vaccine and efficacy trials are in progress. Rolled up lipid bilayers stabilized by calcium bridges containing protein precipitates, termed protein cochleates, stimulate IgG, IgA and CTL against the incorporated protein in mice. It will be interesting to see how this technology develops because it has the potential to incorporate more than one antigen in the formulation.

Another strategy that has been pursued vigorously is to incorporate antigens into solid particles called ISCOMs. A 35 nm cage-like structure is formed between Quil A (the matrix), added lipids and the antigen by mixing a biocompatible detergent with the Quil A and antigen. High levels of antibody and CTL in response to a variety of antigens has been demonstrated using ISCOMs in animals but as yet no human trials have been conducted, at least partly because concerns over the safety of Quil A have yet to be resolved. Difficulties for liposomes, protein cochleates and ISCOMs include the standardization of the formulations to ensure minimal batch-to-batch variation, the stability of the product and scaling up procedures for manufacture.

Biodegradable polymers such as poly(D,L-lactide-co-glycolide) (PL-DLG), used in soluble sutures, are known for their safe use in humans and can be used to encapsulate antigen in microparticles. Ovalbumin, a poor antigen, entrapped in 5 μm PLDLG particles injected subcutaneously in mice, induced a significantly higher primary IgG response than ovalbumin in Freund's complete adjuvant. Microparticles between 5 and 10 μm, containing toxoids, administered orally, are taken up by Peyers patches in the gastrointestinal tract and primed and boosted anti-toxoid IgG and IgA responses in mice. Different formulations of polymer can be used to allow variations in the delivery of antigen. A pulse of antigen could be followed by a trickle of antigen, allowing manipulation of the delivery profile of the antigen. Further clinical studies will be required to establish the efficacy of this approach and determine if low doses will generate unwanted side effects such as tolerance or hypersensitivity.

### 7.6.3 Presentation to the mucosal immune system

Currently most vaccines are delivered by injection, which requires qualified personnel to give repeated immunizations. Oral vaccination, if available, could be administered without the cost of needles, syringes and training of personnel. This would be particularly advantageous in developing countries. As a consequence, a significant amount of effort is being made to develop oral vaccinations. Many infections are acquired at the mucosal surfaces of the respiratory, genito-urinary or gastrointestinal tracts. The primary immune defense mechanism at these mucosal surfaces, IgA, requires presentation and processing of antigen by cells at these sites. Interestingly, there is some evidence that stimulation at one surface can be passed on to distal mucosal sites via trafficing immune cells. This means that stimulation in the gut can lead to local immunity at the lungs and genito-urinary tract. This *common mucosal immune system* increases the attraction of oral immunization. Oral immunization has been attempted with live vectors such as *Salmonella* (see Section 8.8), live attenuated cholera strains (see Section 8.2) and poliovirus (see Sections 8.3 and 8.9) as well as by linking protein to cholera toxin B subunit, either as a gene fusion or by conjugation.

Immunization at other mucosal surfaces may also be possible. SIV p27 *gag*–TY virus-like particles conjugated to cholera toxin B subunit and administered to male monkeys by topical urethral immunization induced sIgA and IgG antibodies to p27 in urethral secretions, urine and seminal fluid.

It should also be noted that although injected antigen stimulates mucosal immunity relatively poorly, it may not be required for protection if disease results mainly from further, systemic spread of the pathogen. Both IPV and OPV prevent the sequelae of poliovirus infection.

## References

1. Kriegler, M. (1990) in *Gene Transfer and Expression: a Laboratory Manual* Stockton Press, New York, pp. 103–112.
2. Takehara, K., Ireland, D. and Bishop, D.H.L. (1988) *J. Gen. Virol.,* **69,** 2763.
3. Mililch, D.R. (1993) in *Hepatitis B Vaccines in Clinical Practice* (R.W. Ellis, ed.). Marcel Dekker, New York, p. 351.
4. Alonson, P.L., Armstrong Scellenberg, J.R.M., Masanja, H. *et al.* (1994) *Lancet,* **344,** 1175.

# Chapter 8

# New and improved live attenuated vaccines

In the past, live attenuated vaccines have been derived quite empirically, usually after extended passage on artificial medium or in tissue culture. Modern techniques now allow the analysis of a pathogen's virulence and antigenicity at the molecular level, and this enables a more rational approach to the generation of attenuated organisms. The ultimate aim is to engineer live vaccines with such desirable properties as total safety and potent antigenicity. In some cases where the DNA is infectious, for example in HSV, the manipulation can be done directly. In other cases, plasmids are used to transfer or delete genetic information by recombination into the genome. The engineering of RNA viruses has proved more difficult, although there has been some success with poliovirus (see Sections 8.3 and 8.9) and with influenza virus. Genetic modification of bacteria has been relatively straightforward where the pathogen grows well but, in the case of slow-growing organisms, such as mycobacteria, it has proved to be more problematical.

It should be noted that, as the number of immunosuppressed individuals increases due to the AIDS pandemic, the frequency of vaccine-associated complications from live vaccines is also likely to increase. Vaccine manufacturers are by nature very cautious, particularly when litigation is possible. It is likely, therefore, that live vaccines will be favored only under very well-defined circumstances, for example if they are markedly superior to anything else available.

Homologous recombination occurs in a variety of systems and has been utilized for the construction of many of the recombinant organisms described in this chapter. It is essentially breakage of two molecules at identical regions of DNA followed by exchange of fragments and religation of the break points. The result is exchange of DNA sequences between molecules (see *Figure 8.1a*). This phenomenon can be used to engineer either deletions or insertions in a variety of circumstances. *Figure 8.1b* shows how a deletion can be generated in the genome of a virus by introducing a plasmid construct and the virus into a cell. The use of a plasmid to delete a region of the chromosome of *V. cholerae* could just as easily have been used to illustrate the process. The recombination process probably proceeds via a single recombination followed by resolution of an unstable intermediate that contains direct repeats. In the figure a double cross-over event is indicated but, whatever the mechanism, the outcome is the same. Recombination is relatively inefficient, and selection procedures are required for differentiation of recombinants from parental genomes.

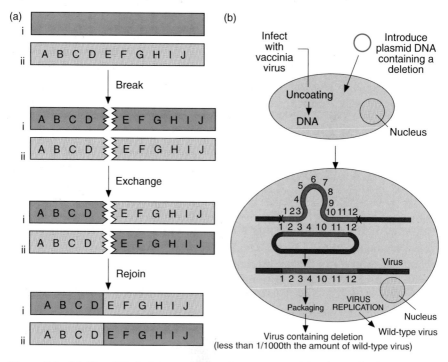

**Figure 8.1:** (a) Simplified picture of homologous recombination. The letters correspond to the DNA sequences. The actual molecular mechanisms of homologous recombination comprise a discrete number of steps that lead to the final outcome. The two sets of sequences (i) and (ii) could be chromosomes or a virus or a plasmid in a cultured cell. (b) Schematic representation of the generation of a vaccinia virus mutant containing a deletion. A plasmid with the desired deletion is constructed and introduced into vaccinia virus-infected cells by transfection. Homologous recombination occurs at low frequency between the plasmid and the virus genome. The resulting DNA is packaged and a virus formed. The virus will be infectious if the sequence deleted is nonessential for growth in cell culture. Although vaccinia virus is illustrated here, the process is equally applicable to other viruses or bacteria. In the case of bacteria, the plasmid is introduced and homologous recombination allowed to occur. It is also possible to introduce foreign DNA with the deletion or interruption of host sequences and generate recombinant viruses expressing foreign genes.

In the case of vaccinia virus, recombinants have been selected with dominant selectable markers, for example *E. coli* guanine phosphoribosyl transferase which confers mycophenolic acid resistance, with indicator genes such as *E. coli* β-galactosidase which permit recombinant virus plaques to be identified by their blue color in the presence of a chromogenic substrate and with a variety of other biological procedures. The gene to be expressed in a recombinant is put under appropriate promoter control and placed adjacent to the selectable marker. Recombinants that are chosen on the basis of the selectable marker should also co-express the gene of interest.

*Figure 8.2* shows a vaccinia recombinant constructed at the Walter–Reed Army Institute of Research in Washington. The parent virus (NYVAC) is a gene-deleted version of the vaccinia virus, vaccine strain Copenhagen. Four sequential homologous recombination events were required to delete the desired vaccinia genes. The result, NYVAC is a highly attenuated vaccinia virus which will not grow in human cells but can be grown on chick cells. A further four sequential homologous recombinations were required in order to insert the seven *P. falciparum* genes shown (each of the seven genes had to be specifically mutated to delete sequences that are seen as early transcriptional termination sites by vaccinia). The recombinant expresses all seven genes in tissue culture and will be tested in phase I/II trials using volunteers during 1994. This example illustrates the power of homologous recombination. It is now relatively easy to manipulate the genomes of many pathogenic agents either to attenuate their virulence or to use them as vehicles for foreign gene expression. Both approaches have the potential to generate new vaccines.

## 8.1 Improving attenuation: pseudorabies – a veterinary vaccine paradigm

In January 1986 the first live viral vaccine on the open market produced by recombinant DNA technology was licensed in the USA. This veterinary vaccine is a molecularly attenuated strain of pseudorabies virus (PrV), a member of the herpesvirus family. It is used to combat the disease caused by the wild-type virus in pigs, which is a serious problem for swine husbandry. Infected piglets may die, and some reproductive problems and increased secondary respiratory disease can occur. Vaccination is used to protect against overt signs of clinical disease but virus replication and excretion, although reduced, are not prevented. Vaccines will also limit virus replication after reactivation. The recombinant vaccine was generated by introducing a plasmid with a 148 bp deletion in the PrV thymidine kinase (TK) gene into cultured cells infected with PrV. Homologous recombination takes place at a low level between PrV genomic DNA at the TK locus and the deleted TK gene carried by the plasmid. The recombinant virus is TK-negative and can be

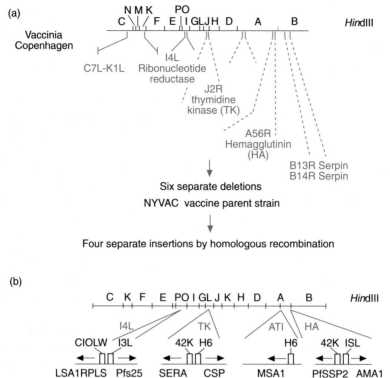

**Figure 8.2:** The construction of a vaccinia virus expressing seven different *P. falciparum* genes. (a) The physical map for *Hin*dIII fragments of vaccinia virus strain Copenhagen is shown. Fragments are identified alphabetically in order of decreasing size. The location of the six separate deletions (see *Figure 8.16b*) used to generate NYVAC are shown. The open reading frames affected are shown as *Hin*dIII fragment (e.g. C), followed by a number and either L for left-reading or R for right-reading. Thus C7L is the seventh left-ward open reading frame in *Hin*dIII fragment C. (b) The highly attenuated NYVAC was used as the parent vaccinia virus to insert seven *P. falciparum* genes (SSP2, Pfs25, SERA, CSP, MSA1, PfsSP2, AMA1) from different stages of the malarial life cycle by homologous recombination. 42K, I3L, H6 and C10LW are vaccinia virus promoters. I4L, TK, ATI and HA are the vaccinia genes for ribonucleotide reductase, thymidine kinase, A type inclusion bodies and hemagglutinin, respectively.

distinguished from parental virus by its ability to grow unimpaired in the presence of 5-bromodeoxyuridine. The TK gene lesion reduces the ability of the virus to establish neuronal replication leading to a latent infection. Regulatory authorities were satisfied that the vaccine was safe, partly because it was based on a strain of virus that had already been attenuated by passage *in vitro*. Since the initial gene-deleted viruses were described, other engineered pseudorabies viruses based on wild-type or attenuated strains have been licensed as vaccines in both the USA and Europe. These viruses have deletions in one or more of three genes associated with virulence: the TK gene and those encoding glycoprotein gI or glycoprotein gX. Some attenuated viruses currently under evaluation carry all three deletions and have a further deletion in the inverted repeat region that also affects virulence.

Use of live attenuated vaccines can lead to problems in diagnosis of natural infections and identification of the cause of an outbreak. Could the outbreak be due to a vaccine strain that has reverted to virulence? This issue was difficult to resolve before the use of the gene-deleted viruses. Serological assays have now been developed that are specific for the wild-type protein which is absent from the vaccine strain. Thus, vaccine strains will not produce antibodies against the deleted gene and can be distinguished from wild-type strains. A further sophistication in some vaccines has been the incorporation of added DNA sequences which have been used to trace and distinguish vaccine virus strains. Such markers in any recombinant vaccine would be useful for clear identification of the vaccines should they be suspected of circulating in the environment or reverting to virulence. Recombination of live vaccines with field viruses could also be followed if modified viruses are genetically marked in this way.

Despite these advances there remains a need to develop more potent vaccines which afford complete protection and prevent infection. In a variety of systems it has been observed that vaccine strains which are completely avirulent lose much of their capacity to replicate and spread within the vaccinated animal. Consequently, much of their immunogenicity is also lost. The hope with genetically engineered strains is that they will be more specifically attenuated and biologically stable than the vaccines used previously.

## 8.2 Improving attenuation: *V. cholerae*

The seventh cholera pandemic, which started in 1961, is now widespread in the Far East, Africa and South America and is caused by the El Tor biotype of cholera. It is estimated that over a million deaths have occurred from this biotype in South America alone since cholera was introduced into Peru in 1991. A further disturbing discovery was the emergence of a new serotype of *V. cholerae*. Both classical and El Tor biotypes of cholera produce a characteristic LPS antigen termed 01, which is known to be a major determinant in inducing protective immunity. In October 1992 a massive cholera epidemic in south Asia was caused by a serotype with a new LPS antigen, 0139 or 'Bengal'. The 0139 serotype is closely related to the El Tor biotype but has acquired new genetic information, and epidemiological data suggest that prior immunity to 01 serotypes will not protect against 0139 strains.

The currently available cholera vaccine is a heat-inactivated, phenol-preserved mixture of the Inaba and Ogawa serotypes. It gives at most a

50% protection for 1–3 months after immunization against both classical and El Tor biotypes. This poor record emphasizes the need for a safe, longer-lasting, vaccine which will not only prevent disease but will also prevent asymptomatic carriage.

It is now accepted that vaccines against enteric infections must be able to stimulate the gut lymphoid tissue to be most effective and that this goal is usually better achieved by administering immunogens orally rather than parenterally. Some progress has been made with two inactivated oral cholera vaccines. These consist of inactivated *V. cholerae* 01 bacteria alone or in combination with purified B subunit of cholera toxin. A randomized placebo-controlled field trial involving over 60 000 individuals in rural Bangladesh showed that the combined B subunit plus inactivated whole-cell (BS-WC) vaccine was safe and effective. Two or three doses conferred 85% protection against cholera for the first 6 months in all age groups tested. After 3 years, protection fell to 51%, with immunity waning most rapidly in the younger age groups. Surprisingly, for the first 3 months of the trial, the BS-WC vaccine was shown to give 75% protection against diarrhea caused by *E. coli* strains that produce an enterotoxin immunologically cross-reactive with the cholera B subunit. It should be possible to adapt this type of vaccine to new serotypes of *V. cholerae* as they arise. Trials will shortly be carried out with an 0139 serotype included in the BS-WC vaccine. The main drawbacks to this approach are that the vaccine is less effective against El Tor biotypes and does not block their carriage in the intestine. It also gives reduced protection in individuals with the O blood group.

Much effort has also gone into the design and engineering of live attenuated cholera strains as they may offer some advantages over the treatments available.

Although virulent strains of cholera may possess a number of virulence determinants, the clinical manifestations of cholera are primarily due to the action of the holotoxin on the intestinal epithelia. Cholera toxin comprises two types of subunits, A and B, encoded by genes located on the bacterial chromosomes. The B subunit specifically binds to the GM1 ganglioside located on the surface of eukaryotic cells, whereas the A subunit possesses ADP-ribosylating activity which results in the ribosylation of the $G_s$ protein of the adenylate cyclase system in host cells. Virulent strains of *V. cholerae* are noninvasive and, therefore, stimulation of systemic immunity by the currently available, inactivated *Vibrio* strains is likely to provide only limited protection against the disease. Because the B subunit of the holotoxin is immunogenic, it has been investigated as a vaccine candidate to stimulate mucosal anti-toxin immunity at the site of infection. Such local production of specific immunoglobulins may lead to inhibition of cholera toxin activity by blocking its adsorption to target cells.

Initial attempts to engineer *V. cholerae* based on a marker exchange protocol proved highly successful (see *Figure 8.3*). A broad host range

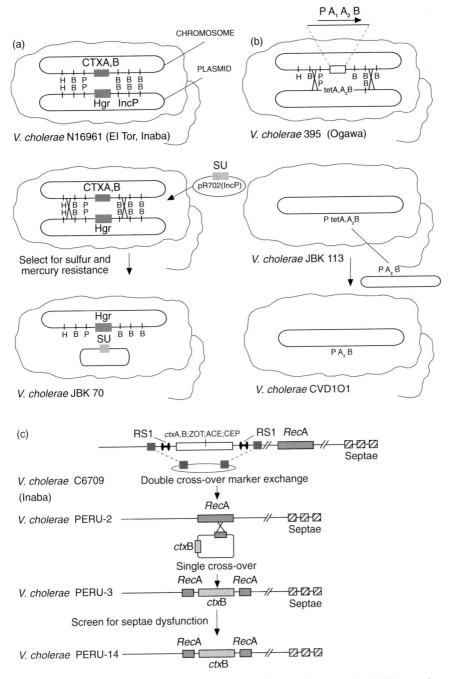

**Figure 8.3:** Generation of *V. cholerae* mutants (for detail, see text). (a) Generation of JBK70; (b) generation of CVD101. CVD103, an analogous mutant of the Inaba strain 569B, and CVD103 HgR, a mercury-resistant derivative, have been tested extensively in clinical trials. (c) Generation of *V. cholerae* defective in recombination.

plasmid belonging to the P incompatibility (Inc P) group, containing *V. cholerae* DNA toxin A and B subunits replaced with a mercury resistance gene, was mobilized into *V. cholerae*. Homologous recombination occurred, with the net result of transferring the mercury resistance to the host cholera genome while deleting the toxin genes. The recombination event was detected by transferring another Inc P group plasmid containing sulfur resistance and selecting for sulfur and mercury resistance. The resulting *V. cholerae* strain JBK70 did not produce either the A or B subunits of cholera toxin. A further strain, CVD101, was also engineered which produced the B subunit but not the A subunit. A single oral dose of JBK70 or CVD101 conferred 89% protection against diarrheal disease in volunteers challenged 1 month later with a virulent El Tor Inaba strain. Both of these vaccine strains displayed similar 'reactogenicity' and gave mild to moderate diarrhea in slightly more than half the vaccinees.

CVD103-HgR is a *ctx*A deletion derivative of the classical biotype strain 569B and has reduced reactogenicity. CVD103-HgR has been shown to be well tolerated and highly immunogenic both in adults and in children following administration of just a single oral dose; a large-scale field trial in Indonesia of a lyophilized oral formulation is underway.

Virulence determinants, other than the holotoxin, that have been implicated in mediating infection by cholera vibrios include colonization antigens and secondary toxins. New oral vaccine candidates, such as *V. cholerae* CVD110 El Tor, Ogawa, have all known virulence genes deleted. Three of those encoding cholera toxin (*ctx*), zonula occludens toxin (*zot*), and accessory cholera enterotoxin (*ace*), are located on a 4.5 kb virulence cassette flanked by repetitive sequences (RS1 elements). Homologous recombination between these RS1 elements resulted in the deletion of this virulence cassette to yield *V. cholerae* CVD109. Insertion of genes encoding mercury resistance (*HgR*) and the cholera toxin B subunit (*ctx*B) into the hemolysin locus (*hly*A) produced CVD110. This insertion serves three purposes:

(1) it genetically tags the vaccine strain to distinguish it from wild-type *V. cholerae* 01;
(2) it produces cholera toxin B subunit in order to elicit antitoxin immunity; and
(3) it inactivates the hemolysin gene, rendering the strain nonhemolytic on sheep erythrocyte agar plates.

The efficacy of this type of strain is now being tested in volunteers. Other workers have focused on the risk of attenuated strains reacquiring the virulence genes. Countermeasures have included a deletion in the *Rec*A gene of *V. cholerae*, thereby inactivating site-specific homologous recombination in the host *Vibrio*. Peru-3, Peru-14, Bengal-3 and Bengal-15 are oral vaccine candidates. The toxin genes located on the RS1 element have been removed from each and all have a deletion in the *Rec*A gene. Peru-3 and -14 are based on the 01 serotype and Bengal-3 and -15 are based on the 0139 serotype. Peru-14 is a filamentous mutant of Peru-3 and is nonmotile. Bengal-15 is also a nonmotile mutant of Bengal-3. Both Peru-14 and Bengal-15 give over 80% protection and are less reactogenic than their isogenic counterparts.

Thus a variety of approaches to attenuation are being taken and it is likely that within the next few years more effective vaccines for cholera will become available. It is undoubtedly true that the most effective measure to prevent cholera would be to improve sanitation and provide clean drinking water but the costs associated with this are astronomical and it is likely that an effective vaccine would be used for many years to come.

## 8.3 Improving stability: poliovirus

The WHO has called for improvements to the OPV, particularly the type 3 virus component. One of the ways of doing this is to engineer more stable attenuated viruses. The technique for producing engineered polioviruses stems from the observation that infectious viruses can be rescued from cDNA clones when the appropriate sequence is put under the control of a strong eukaryotic promoter and transfected into susceptible cells. Thus, by standard molecular technology, it is possible to introduce defined mutations or alterations into the cDNA and isolate a recombinant virus (see *Figure 8.4*).

The three Sabin attenuated strains of poliovirus have been used successfully for many years to protect against paralytic poliomyelitis. These strains are very stable but, at low frequency, it has been shown that the type 2 and 3 strains can revert to virulence and cause vaccine-associated disease. The Sabin type 1 strain, however, is much more stable. Two possible approaches to the generation of more stable type 2 and type 3 viruses have been investigated. The first is to replace the type 1 major virus neutralizing epitopes with type 3 epitopes (as in Section 8.9). This would possibly give a virus as stable as the Sabin type 1 strain but with the major antigenic characteristics of type 3. A second approach was based on the observation in a monkey model that a strong neurovirulence determinant resided in the 5'-noncoding sequence of the genome and that the surface structure of the virion particle had less correlation with the neurovirulence or attenuation phenotype. The 5'-noncoding region of the type 1 strain was used in place of the 5'-noncoding region of the type 3 strain. Here the chimeric virus RNA codes for a completely type 3 virus particle but it is hoped that it will have a more stable phenotype. A further improved phenotype that might be achieved by the introduction of mutations is an increase in the thermostability of the type 3 vaccine.

These partially theoretical improvements may still not be taken up by vaccine manufacturers for a variety of reasons, not the least of which is

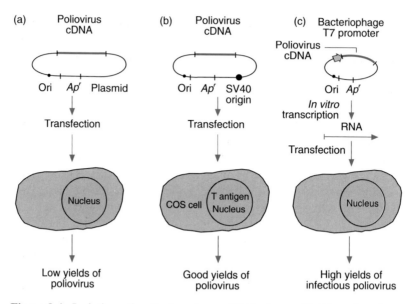

**Figure 8.4:** Isolation of poliovirus from cDNA clones. Full-length poliovirus cDNA, when transfected into cells, will give low yields of infectious virus (a). If an SV40 origin of replication is incorporated into the plasmid DNA and the construct is transfected into COS cells (b), yields of virus are dramatically improved due to replication of the input plasmid mediated by T antigen produced by the COS cell. A second refinement is to produce an RNA *in vitro* that can be used as a template to generate poliovirus proteins. This is achieved by using T7 RNA polymerase and a plasmid that contains the full-length cDNA clone adjacent to a T7 bacteriophage promoter. $Ap^r$, ampicillin resistance gene; Ori, bacterial origin of replication.

the difficulty of showing conclusively that a complication rate in humans is reduced from 1 in $10^6$ vaccinations to an even smaller number!

## 8.4 Recombinant live vectors

This strategy is based on the use of currently available live attenuated vaccines as hosts for foreign genes. A gene of choice coding for a protective antigen is expressed by the live vaccine. Upon vaccination with the recombinant, immune responses to the carrier vaccine and to the foreign gene are generated. It is hoped that they are sufficient to elicit immune responses both against the pathogen from which the foreign gene is derived and against the parent vaccine.

## 8.5 Vaccinia virus recombinants

The use of recombinant live vectors has been pioneered with vaccinia virus. In 1984 a vaccinia virus expressing the HBsAg was shown to protect immunized chimpanzees against liver damage normally associated with challenge with HBV. Over the intervening 10 years, phase I clinical trials in humans and field trials of a rabies virus vaccine for immunizing wildlife have confirmed the potential of these vectors, although safety issues continue to cause concern.

Vaccinia virus was used for over 100 years as a live attenuated vaccine for the control of smallpox. The vaccine is easy to mass-produce at low cost; in the smallpox eradication campaign a single dose of freeze-dried vaccine cost less than US¢2. Although using current tissue culture techniques manufacture would cost more, it would be much cheaper than the current price of the hepatitis B vaccine which is prohibitively expensive for many developing countries. The ability to freeze-dry the smallpox vaccine means the virus is very stable for long periods at high temperatures and eliminates the need for costly and difficult cold chains to be set up. Its potency as a single inoculation and stimulation of both cell-mediated and antibody responses, as well as ease of administration under field conditions, are all advantages associated with the use of vaccinia in the smallpox eradication campaign. These advantages should also be benefits associated with the use of recombinant vaccinia viruses that express foreign genes. Over 150 different vaccinia recombinants expressing genes from viral, bacterial and parasitic pathogens have been described. Many have been shown to protect systems against challenge with the appropriate pathogen in animal models. *Table 8.1* shows progress in the trial of several recombinant poxvirus vaccines being tested in humans. Antibody responses to the foreign gene have been somewhat disappointing, particularly in individuals previously immunized against smallpox. However, it was noted in an animal model that a boost with the recombinant HIV-1 glycoprotein gp160 after immunization with a vaccinia recombinant expressing gp160 gave higher levels of antibody than were raised with either the vaccinia recombinant virus or the recombinant glycoprotein alone. Several of the trials have incorporated this information into their protocol and have included boosts with HIV-1 gp160.

Plans for use of two vaccinia-based vaccines in animals are also well advanced; one to protect cattle against rinderpest, the other to protect wildlife against rabies. The closest to extensive use is the vaccinia recombinant expressing the rabies virus glycoprotein. It has been shown to induce neutralizing antibody and CTL in vaccinated animals. More impressively, it protects foxes, fox cubs, skunks and raccoons against challenge with the wild-type rabies virus even when the vaccine is presented as baited food.

Three vaccination campaigns were carried out over an area of $2200 \text{ km}^2$ in Belgium in 1989 and 1990 by dropping 25 000 machine-made baits containing $10^8$ $TCID_{50}$ of a vaccinia recombinant that expresses the rabies virus G glycoprotein (*Figure 8.5*). In France similar baits were dropped by helicopters over an area of $10 000 \text{ km}^2$. In both cases the density was 15 baits $\text{km}^{-2}$. In addition to an attractant for foxes, the baits contained tetracycline and the recombinant vaccinia virus. Tetracycline

**Table 8.1:** Human vaccination with poxvirus recombinants

| Virus/recombinant gene | Immunization procedure | Immune response | | Outcome |
|---|---|---|---|---|
| | | Humoral | Cell-mediated | |
| Vaccinia strain Tian Tan expressing EBV gp340 from 11 kDa late structural promoter | All scarified. 11 vaccinia and EBV-positive adults (A)–108 p.f.u. ml$^{-1}$; 6 EBV-positive, vaccinia-negative schoolchildren (B)–107 p.f.u. ml$^{-1}$; 9 children under 3 (C)–107 p.f.u. ml$^{-1}$ | (A) no response to gp340; (B) anti-gp340 titers boosted; (C) children develop anti-gp340 titers | ND | Partial efficacy (reported at meetings) |
| Vaccinia WR expressing HIV-1 gp160 from 7.5 kDa promoter | Scarified followed by boosting with fixed autologous vaccinia gp160 recombinant infected cells | Neutralizing antibody | CTL activity | Antibody and T-cell response |
| Vaccinia WR expressing HIV-1 gp160 from the 7.5 kDa promoter | See detail in [1]: boost with fixed vaccinia gp160 infected autologous PBL | Transient and weak | Long-term memory | Two-year follow-up indicates CTLs produced |
| Vaccinia Wyeth expressing HIV-1 gp160 from the 7.5 kDa promoter | See ref. [2] for details | 2/31 | 13/16 proliferate to soluble gp160 | Weak T cell-specific responses |
| As above boosted by purified gp160 | See ref. [3] for details | Neutralizing antibody in 7/13 | Good proliferative responses to gp160 | Sustained high levels of antibody in 75% of recipients for > 18 months |
| As above boosted by 640 μg baculovirus-derived gp160 | See ref. [4] | Neutralizing antibody in 8/12 (3/12 cross-neutralizing) | ND | High levels of antibody (higher than either vaccinia recombinant or protein on its own) |
| Canarypox expressing the rabies virus glycoprotein | See Section 8.6 and ref. [7] for details | + | ND | 9/9 high-dose recipients produced good levels of rabies virus neutralizing antibody |

ND, not determined.

**Figure 8.5:** Field trials of fox vaccination with the VVTGgRAB in the province of Luxembourg (South Belgium). Baiting areas: (1) military field of Marche-en-Famenne (6 km$^2$, September 1987); (2) region of Neufchâteau (435 km$^2$, October 1988); (3) half part of the province of Luxembourg (2200 km$^2$, November 1989, April and October 1990). Reproduced from ref. [5]. Reprinted by permission of CRC Press, Boca Raton, FL.

can be detected easily by its fluorescence under UV light in the bones of any animal eating the bait, so it provides a useful biomarker of bait uptake. Between 80 and 96% of the baits dropped were consumed within 30 days. Adult foxes, cats, dogs, wild carnivores and rodents were then collected from within the treated area and assessed at post mortem for uptake of tetracycline. After the second campaign, 62% of the foxes examined were positive for the biomarker and a number of other species were also positive for tetracycline. Long-term monitoring of the target areas has not revealed any problem with spread of the virus and laboratory-based infection of a variety of indigenous species from voles to carrion crow indicated that the vaccinia recombinant virus was innocuous. *Figure 8.6* shows that the cases of rabies within wild animals in the target area dropped dramatically in the treated area, despite high densities of foxes in the area in 1990. In France similar results have been obtained with an uptake of the bait in foxes as high as 70%, and preliminary results indicate that more than 80% of the foxes had seroconverted for rabies G protein. The efficacy and heat stability of the vaccinia recombinant offer an excellent alternative to the attenuated strains of rabies virus currently used in the field. Raccoonpox has also been used as a vector for rabies virus G and will protect the raccoon (the major vector for rabies in the USA) against challenge with rabies virus.

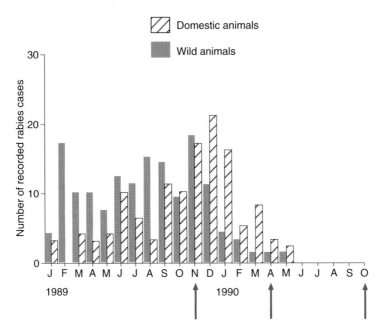

**Figure 8.6:** Number of recorded rabies cases during vaccination campaign. Rabies incidence in domestic and wild animals in a 2200 km$^2$ area in 1989/1990. Arrows indicate the timing of campaigns of fox vaccination using the vaccinia–rabies recombinant. Reproduced from ref. [5]. Reprinted by permission of CRC Press, Boca Raton, FL.

Poxviruses have been found in many species of animal, and they often have a limited host range. Fowlpox virus, for instance, will only replicate in avian species. Thus it is well suited as a vector for expression of vaccine antigens to immunize poultry against pathogens such as Newcastle disease virus (NDV). Indeed, recombinants expressing genes from NDV will protect poultry against the disease (see *Table 8.2* for fowlpox recombinant used). Other poxviruses, such as capripoxviruses and swinepox, are being engineered to express antigens relevant for vaccination of ruminants and pigs, respectively.

## 8.6 Cross-species vaccination: 'live–dead' vaccines

Canarypox virus replicates in cells of avian origin but is blocked in its ability to replicate in human cells. Recombinant canarypox viruses expressing the rabies virus glycoprotein have been tested in phase I clinical trials in humans and shown to induce rabies virus neutralizing antibody. Because canarypox virus cannot replicate in the vaccinee, there is no danger of virus spread. However, the recombinant virus does replicate sufficiently in human cells to produce the foreign gene product. This 'live–dead' vaccine may prove to have the advantages of live vaccines with authentic antigenic presentation and the generation of antibody and CTL

**Table 8.2:** Immunization of poultry with fowlpox recombinants

| Recombinant gene expressed | Challenge virus | Outcome |
| --- | --- | --- |
| Infectious bursal disease virus (IBDV) VP2–β-galactosidase fusion | IBDV | Chickens protected from mortality, not from damage to the bursa |
| Avian influenza HA | Pathogenic avian influenza | Wing web vaccinated totally protected, comb scarified; partially protected from lethal challenge |
| Newcastle disease virus (NDV) | NDV | Protection of chickens |
| HA–NA gene fusion protein | NDV | Protection of chickens |
| Marek's disease virus (MDV) gB or gB and pp38 | MDV | Protection of chickens |
| Avian reticuloendotheliosis retrovirus (ARR) envelope glycoprotein | ARR | Decrease in viremia |

responses without the possible complications associated with live viruses. This principle might also be extended to other viruses, although it is difficult to envisage this approach being viable in most instances because restricted replication and lack of virus spread will undoubtedly limit the immune response to the vector and foreign gene.

## 8.7 BCG

BCG is an avirulent bovine tubercle bacillus that is the most widely used vaccine in the world. Within the last few years it has been possible to engineer BCG as a vector to express antigens from other organisms, for example the envelope glycoprotein of HIV. Recombinant BCG has a number of distinct advantages over other approaches to multivalent vaccines, primarily the experience gained with the parent BCG vaccine. Other advantages include the fact that BCG is one of only two vaccines (the other is OPV) which the WHO recommend to be given at birth. The younger the age at which vaccination can begin, the better the chances of success in vaccination programs. A single immunization with BCG can give long-lasting cell-mediated immunity to tuberculosis; it can be given repeatedly; complications are rare; and BCG is a highly potent adjuvant in its own right. Although bacteriophage and plasmid vectors have been used with some success to construct BCG recombinants, a significant amount of development is still required both to achieve higher levels of expression and to allow the system to be more readily manipulated (see also Section 11.5).

## 8.8 Attenuated salmonella strains as live recombinant bacterial vaccines

It is possible to introduce totally defined mutations or deletions in a variety of bacterial strains in order to achieve attenuation. These rationally designed attenuated vaccines can also be used as carriers for antigens cloned from other pathogenic organisms. Attenuated salmonella strains seem good candidates for this approach, partly because of experience gained from the widespread use of the live attenuated TY21a strain as a vaccine for *S. typhi* and partly because oral administration stimulates secretory antibody and cellular immune responses in the host. For example, the gene for heat-labile B subunit of enterotoxic *E. coli* was introduced into the attenuated *AroA* strain of salmonella. This recombinant salmonella was able to induce IgG and IgA antibodies to the enterotoxic B subunit (as well as against salmonella) in vaccinated animals. A further modification of this strategy is to incorporate peptides into the flagellin gene of salmonella. A potential HBV vaccine was constructed by incorporating synthetic oligonucleotides coding for sequences from the HBsAg and from the preS2 antigen into the flagellin gene followed by introducing the hybrid gene into a flagellin-negative salmonella strain. The recombinant salmonella expressed the hybrid flagellin gene and, when used to vaccinate mice, guinea pigs or rabbits, induced antibodies that recognized native HBsAg. In addition, T lymphocytes isolated from immunized mice proliferated in response to the hepatitis peptide contained in the flagellin gene, showing that the cell-mediated immune response can also be generated by recombinant salmonella. The antibody responses in mice immunized by i.m. inoculation were greater than those vaccinated by the oral route.

Despite these successes, continued efforts to improve oral immunization are important because of the ease of administration and reduced costs associated with this route (syringes and needles are not required). Also, mucosal immune responses to antigens after oral vaccination may offer more protection against pathogens that have their initial replicative cycle on similar mucosal surfaces. If this approach proves particularly successful, it may be possible to vaccinate with multivalent recombinant salmonella strains that contain genes from rotaviruses, *Shigella* spp. and other pathogens which infect the gastrointestinal tract.

## 8.9 Poliovirus chimeras

The live attenuated poliovirus type 1 Sabin strain has proved to be a very safe and effective vaccine stimulating good antibody responses. A knowledge of the crystal structure of the virus (*Figure 8.7*), together with the ability to generate virus from cDNA molecules (see Section 8.3), has allowed antigenic domains from other pathogens to be incorporated precisely into the virus particle at the most antigenic sites.

Trachoma and STDs caused by *C. trachomatis* are major health problems worldwide. Epitopes on the major outer membrane protein (MOMP) of *C. trachomatis* have been identified as important targets for the development of vaccines. A poliovirus hybrid was constructed in

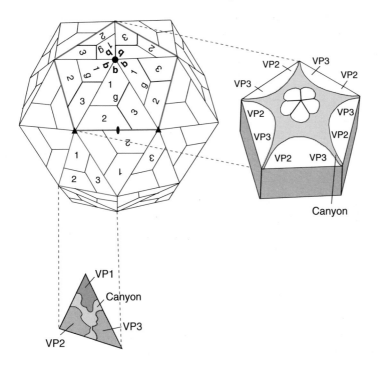

**Figure 8.7:** Diagrammatic representation of the structure of poliovirus virions and major antigenic sites. Some of the copies of VP1, VP2 and VP3 are labeled (1, 2 and 3, respectively). The letters 'b' and 'g' show positions of the BC and GH loops in VP1. The virus particle has prominent 'dimples' on its fivefold axis of symmetry and a flexible protrusion near the positions marked 'g'. Reproduced from ref. [6] with permission from Raven Press.

which part of antigenic site I (a poliovirus neutralizing epitope) of Sabin poliovirus type 1 was replaced by a sequence from variable domain I of the MOMP of *C. trachomatis* serovar A. The chlamydial sequence included the neutralization epitope VAGLEK. This hybrid was viable, grew very well compared with the parent virus and expressed both poliovirus and chlamydial antigenic determinants. When inoculated into rabbits, it was highly immunogenic, with anti-chlamydial titers 10- to 100-fold higher than those induced by equimolar amounts of either purified MOMP or a synthetic peptide expressing the VAGLEK epitope. This antibody was strongly neutralizing *in vitro* and *in vivo*. Because poliovirus infection induces a strong mucosal immune response in primates and humans, these results indicate that poliovirus–chlamydia hybrids could become powerful tools for the study of mucosal immunity to chlamydial infection and for the development of recombinant chlamydial vaccines. Other examples of hybrid polioviruses are several expressing HIV-1 gp160 V3 loop sequences; these can induce HIV-1 neutralizing antibody in immunized animals.

Poliovirus chimeras suffer some of the same limitations as peptide vaccines, and many antigenic sites cannot be incorporated because nonviable virus is produced.

## 8.10 Other virus vectors

A number of other virus vectors which are appropriate for vaccination have been described in the past few years. These include both replication-deficient and replication-competent recombinant adenoviruses. Adenovirus types 4 and 7 have been used to immunize military personnel for some time and, because it is now possible to engineer these viruses to express vaccine antigens, they have some potential as live recombinant vaccines. Poliovirus, HIV and HBV antigens have been expressed in these viruses and such recombinants can induce significant levels of antibodies to the foreign gene products. At present there are reservations about these viruses, particularly due to the large amount of virus needed to achieve effective vaccination. For replication-competent viruses there are concerns over the E1a and E1b gene products which have oncogenic potential, and for the replication-defective variants there is concern over the cell line that is required for growth of the recombinants. A recent clinical trial of an adenovirus type 7 recombinant expressing HBsAg suggests that antigen expression or presentation may not be sufficient to elicit protective immunity.

Although RNA viruses are more difficult to modify, both influenza A virus and poliovirus can now be engineered to express foreign proteins. (Sections 8.3 and 8.9 refer to the engineering of poliovirus). Recently, 400 amino acids of several different foreign proteins have been expressed by poliovirus recombinants. This is done essentially by adding a coding sequence at the 5' end of the virus cDNA. The construct is engineered to express a polyprotein with a 5'-foreign coding sequence adjacent to a synthetic poliovirus protease 3c cleavage site. Replication of virus occurs and the protease produced from the polyprotein cleaves off the foreign gene. Both influenza A and polioviruses have expressed a variety of genes and it will be interesting to see if they offer distinct advantages over other live vectors.

The Oka strain of varicella-zoster virus (VZV; chickenpox virus) is a vaccine licensed in many countries which can be engineered to express foreign antigens. For example, HBsAg has been expressed and immunization with the recombinant was shown to induce antibody to HBsAg.

HSV type 1 has also been used as a vector for several candidate vaccine antigens. In the original reports the foreign gene was inserted into the TK site of the virus resulting in a TK-negative phenotype. TK⁻ viruses are attenuated compared to wild-type virus and will not reactivate from latency, but also can not be treated with anti-herpesvirus drugs such as acyclovir. A new strategy has been described recently which involves the generation of a genetically disabled virus. A cell line expressing the essential HSV glycoprotein, gH, has been constructed. After the virus gH gene is deleted from HSV by homologous recombination the variant can grow only on that cell line. A virus grown on this cell line contains gH in its envelope and can infect cells normally, but a progeny virus is unable to spread from infected cells because gH is required for egress of virus from the cell. Recombinant HSV that are gH negative and express vaccine antigens can then be constructed and used as vaccines. One round of replication will occur, and the infection will abort. These so-called disabled infection single cycle (DISC) vaccines combine some of the advantages of live vaccines with the safety of inactivated vaccines. Apart from their potential to carry foreign antigens they should also be highly effective against HSV infections. Other herpesviruses, such as cytomegalovirus (CMV) and VZV, could be engineered in an analogous way to express foreign genes or as vaccines for CMV and VZV.

## 8.11 Recombinant *E. coli* strains

Enterotoxigenic *E. coli* (ETEC) strains cause diarrheal diseases in young pigs and, under some circumstances, in man. These bacteria adhere to the intestine of the host via surface-associated fimbriae and secrete toxins which can be classified into heat-stable (ST-toxin) and heat-labile (LT-toxin). The fimbriae are highly antigenic, and the first vaccines against ETEC consisted of whole cells or acellular extracts enriched for fimbriae.

Vaccines prepared from ETEC strains gave significant adverse reactions due to high levels of LPS and capsular antigens on the surface of the wild-type ETEC strains. *E. coli* K12 gave far fewer adverse reactions and was used as a vector for plasmid constructs that expressed one or more different antigenic types of fimbriae. However, to produce a vaccine with a wider spectrum of protection, an anti-toxin component was introduced. Plasmid vectors were constructed using a strong prokaryotic promoter that expressed the LT-toxin B subunit at high levels. The Cetus Corporation now market a pig vaccine which consists of an *E. coli* K12 strain that expresses high levels of the LT-B subunit and contains fimbriae from ETEC strains. This engineered *E. coli* was the first licensed vaccine produced by recombinant DNA technology to be used in the USA.

## 8.12 Improving immunogenicity: examples from vaccinia virus

A number of possible approaches to improving immunogenicity have been analyzed using vaccinia recombinants. Vaccinia virus has many genes which appear to be involved in controlling the host's innate and induced immune responses to the virus. There are three genes involved in evasion of the antiviral effects of IFN-$\gamma$ as well as genes that interfere with complement fixation and processing of antigens. Deletion of any of these genes may result in an increased immunogenicity due to a more vigorous response to the virus. They may also become less attenuated, but this seems not to be the case. Insertion of a foreign gene into the serine protease inhibitor (*serpin*) genes of the vaccinia virus genome improves the immunogenicity of the foreign gene product. This may be due to an absence of the *serpin* gene which, if present, might diminish the immune response by preventing processing of antigens and subsequent presentation in conjunction with MHC class I molecules. It is conceivable that other pathogens have adapted to their host and acquired the ability to modify the host immune response. Identification of these genes and their deletion may improve immune responses in other pathogens.

A second possible route to improved immunogenicity of live vectors is to engineer the organism to express an appropriate cytokine. IL-1, IL-2, IL-4, IL-6 and IL-10 have all been expressed in vaccinia virus and, at least, expression of IL-2 has been shown to attenuate the virus markedly in immunocompromised hosts. The effect of co-expression of IL-12 will be interesting as this interleukin has been shown to have adjuvant activity in its own right.

Another possible approach is to fuse a structural gene of the pathogen with the vaccine antigen under study. This has been done in pilot experiments with vaccinia virus where a marker gene ($\beta$-galactosidase) was fused to a structural gene of the virus. A recombinant virus was generated which expressed this fusion product and it was shown to be part of the virus particle. Such presentation as a structural component of the virion should enhance response to the vaccine antigen. In principle, this may be extended to any organism and, indeed, foreign genes have been incorporated into fimbriae which are expressed on the surface of bacteria such as *E. coli*.

# References

1. Zagury, D., Bernard, J., Cheyner, R. *et al.* (1988) *Nature (Lond.)*, **332**, 728.
2. Cooney, E.L., Collier, A.C., Greenberg, P.D., Coombs, R.W., Zarling, J., Arditi, D.E., Hoffman, M.C., Hu, S.-L. and Corey, L. (1991) *Lancet*, **337**, 567.
3. Cooney, E.L., McElrath, M.J., Corey, L. *et al.* (1993) *Proc. Natl Acad. Sci. USA*, **90**, 1882.
4. Graham, B.S., Belshe, R., Dolin, R. *et al.* (1992) *J. Infect. Dis.*, **166**, 244.
5. Pastoret, P.P., Brochier, B., Blancou, J. *et al.* (1990) in *Recombinant Poxviruses*, 2nd edn (M.M. Binns and G.K. Smith, eds). CRC Press, Boca Raton, p. 198.
6. Fields, B.N. and Knipe, D.M. (1990) *Virology*, 2nd edn. Raven Press, New York, pp. 46, 519 and 521.
7. Cadoz, M., Strady, A., Meignier, B., Taylor, J., Tartaglia, J., Paoletti, E. and Protkin, S. (1992) *Lancet*, **339**, 1429.

Chapter 9

# EBV vaccine development: a laboratory-based case study

## 9.1 Natural history and clinical significance of EBV

EBV (*Figure 9.1*) is a member of the herpesvirus family and infects the human population worldwide. In countries such as England and the USA, 80–90% of all adults carry the virus, whereas rates of infection can be nearly 100% in developing countries. The primary route of transmission is believed to be from small amounts of EBV in the saliva of the seropositive individuals to mucosal surfaces in the oropharynx of the uninfected contact. Childhood infection is generally asymptomatic whereas infection later in adolescence presents clinically as infectious mononucleosis (IM) in about half of those infected. Although as many as 100 000 cases of IM occur each year in the USA alone, causing significant morbidity, interest in EBV has primarily been fueled by the association of the virus with several different tumors (*Table 9.1*).

Endemic Burkitt's lymphoma (BL) occurs predominantly in the malaria belts of central Africa and in New Guinea (*Figure 9.2*) where it affects mostly children, often presenting as a B-cell tumor in the jaw. Various lines of evidence implicate EBV in the etiology of the tumor, including the fact that EBV-specific DNA, RNA and gene products can be found in all biopsies of the lymphoma. Probably the best evidence for a causal association comes from a WHO-sponsored prospective serological study in Uganda where 42 000 children were monitored for antibodies to EBV capsid antigen (VCA). The study showed that children with an anti-VCA titer at least fourfold greater than controls had a 30-fold greater risk of developing BL. Epstein pointed out that this risk factor is greater than that accepted as providing an etiological link between smoking and lung cancer.

Nasopharyngeal carcinoma (NPC) is a serious health problem in southeast Asia, particularly in ethnic Chinese populations. For example,

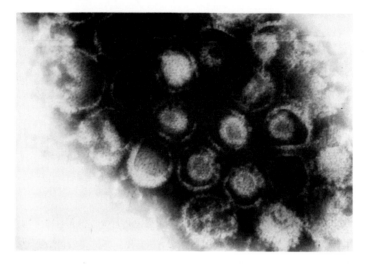

**Figure 9.1:** Electron-microscopic appearance of EBV by negative staining. Virus was concentrated from the supernatant fluid of the B95-8 marmoset cell line.

| | No. of cases 0–14 year olds | 15+ years old |
|---|---|---|
| Lymphoma | 267 | 36 |
| All others | 362 | 5242 |

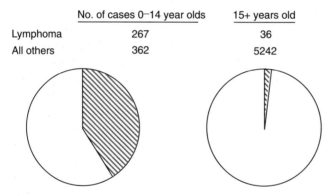

**Figure 9.2:** Burkitt's lymphoma as a percentage of cancers in Uganda (Kampala cancer register 1963–1966).

in Hong Kong males it is the second most common form of tumor, occurring with an incidence of approximately 30 cases per 100 000 individuals per year (see *Figure 9.3*). Although rare in other parts of the world, small areas of high prevalence are found in Alaska, Iceland and in parts of central and north Africa. Worldwide it is estimated that 80 000 cases occur annually, with a peak incidence at 50–60 years of age. There is generally a high mortality rate as diagnosis is often at a late stage when the tumor has already metastasized to the lymphatic system.

It has been suspected for some time that EBV plays an important part in the etiology of NPC. The evidence for this conclusion is based on several findings. EBV DNA, RNA and protein are found in the vast majority of NPC tumors; the form of virus DNA in the tumors is clonal, implying that the virus had infected the original tumor cell; the anti-EBV antibody

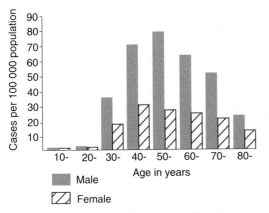

**Figure 9.3:** Age-related incidence of NPC in Hong Kong (1978–1982).

profile (IgA antibody directed against EBV VCA, DNase and TK) is diagnostic of NPC and is altered when tumors are surgically removed; and EBV encodes at least one gene (latent membrane protein, LMP) which can alter the growth properties of epithelial cells in culture.

Relatively recently it has been shown that, in 50% or more of cases, EBV DNA and gene products can be found in Reed–Sternberg cells which are characteristic of Hodgkin's lymphomas (see *Figure 9.4*). Hodgkin's disease has a prevalence of two to four cases per 100 000 individuals in the UK and USA, and this widens the potential of an EBV vaccine. From the perspective of a Western vaccine manufacturer, vaccines for the developing countries have little prospect for financial profit as large investments are required to manufacture pharmaceutical products for human use. The tentative link of EBV with Hodgkin's lymphoma and with a number of T- and B-cell lymphomas has brought the market into the setting of the developed countries.

Although their incidence is lower, other important EBV-associated tumors arise as a result of immune deficiency. In organ-transplant patients receiving imunosuppressive drugs to prevent graft rejection, EBV-driven lymphomas arise with high frequency, and in AIDS patients both EBV-associated lymphomas and hairy oral leukoplakia arise fairly frequently. Children with X-linked lymphoproliferative syndrome (Duncan's disease) fail to mount an immune response against EBV and die as a result of a massive lymphoproliferation. Although the condition is rare, children in families affected by Duncan's syndrome would be one of the first groups to benefit from an EBV vaccine. The consequences of EBV-related diseases are summarized in *Table 9.1*.

## 9.2 Rationale for a vaccine program

It is rather paradoxical that EBV should be involved in tumorigenesis. Although the virus is prevalent in all human populations the incidence of

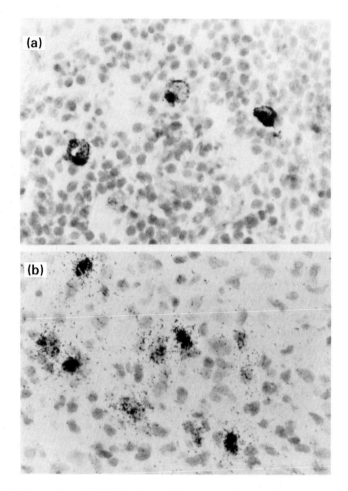

**Figure 9.4:** Detection of EBV in the malignant Reed–Sternberg (HRS) cells of Hodgkin's disease (HD). (a) Immunohistological staining of an EBV-positive HD biopsy with monoclonal antibodies to the LMP1 protein of EBV. Positive staining (arrowed) indicates that in this biopsy LMP1 in expressed. (b) *In situ* RNA–RNA hybridization reveals strong nuclear labeling in tumor cells with EBER-specific (EBV small RNA) probes. H + E counterstained sample was magnified × 340. Reproduced from *The Journal of Experimental Medicine* (1993) Vol. 177, pp. 339–349, by copyright permission of the Rockefeller University Press.

BL and NPC is restricted to certain geographical areas and, even in high prevalence areas, the virus persists asymptomatically in the majority of individuals. Clearly, other factors must be important in the development of these tumors but the weight of evidence suggests that EBV also plays an essential role. It was this close association between EBV, BL and NPC that led to the suggestion that an EBV vaccine may prevent virus infection and the subsequent development of these tumors, as well as preventing morbidity associated with IM. Ultimately, it is the success or failure of a vaccine program that will establish whether or not EBV plays

**Table 9.1:** EBV-related diseases

| Disease | Consequence |
|---|---|
| Infectious mononucleosis | Generally self-limiting/temporarily debilitating |
| Burkitt's lymphoma | Responds well to chemotherapy |
| Nasopharyngeal carcinoma | Low 5 year survival rate, detection of tumor most common after metastasis has occurred |
| Lymphoma | |
|   Duncan's syndrome | Death from massive lymphoproliferation |
|   Transplant | Immunosuppresive drugs allow outgrowth of lymphoma, treated by removing immunosuppression. EBV CTL control lymphoma |
|   AIDS | EBV replication in hairy oroleukoplakia of AIDS patients can be found as well as lymphoma |
|   Hodgkin's | EBV DNA and proteins found in possibly as many as 50% of lymphomas |

a causative role. It should be noted that, although with BL an answer may be available within 10 years of the commencement of a vaccine program, with NPC it is likely to be over 30 years before the question can be resolved.

Although there is considerable variation in the degree of expression of virus genes among different tumors associated with EBV (see *Table 9.4*), it may be possible to control these tumors through vaccination. It may be attempted before transplant to prevent outgrowth of lymphomas after immunosuppression or after transplant to control the outgrowth of the lymphoma. Therapeutic vaccination may also be relevant for BL and NPC.

## 9.3 Vaccine design for EBV

Several important questions must be answered, if only in part, to develop a rational approach to EBV vaccination. These are considered in the following sections.

### 9.3.1 What evidence is there to suggest vaccination against EBV might be effective?

There is very little evidence of individuals being infected with two strains of EBV, and IM does not seem to occur in individuals who are EBV-seropositive. Both facts imply that initial exposure to EBV raises sufficient protection against disease due to later infection. Animal model systems (see Section 9.3.2) also indicate that a vaccine will be effective.

### 9.3.2 Which immunological mechanisms are required for protection against EBV?

In the natural infection, EBV-specific $CD8^+$ CTL control the proliferation of

EBV-bearing B cells. It is not clear if pre-existing CTL would prevent virus latency. It is also not clear whether circulating antibodies to EBV would prevent the initial infection of cells in the nasopharynx. Cottontop tamarins challenged with a large dose of EBV succumb to an EBV-driven lymphoma. However, if they are immunized with the EBV major envelope glycoprotein, gp340/220, they are protected from developing the tumor. Thus anti-gp340 antibody has a protective effect and is likely to be useful in a vaccine.

### 9.3.3  At what anatomical site would the immune response against EBV be most effective?

It is assumed, although not proven, that primary infection with EBV involves virus entry into epithelial cells in the oropharynx. It has not been possible in the laboratory to infect epithelial cells in culture, partly because they lack the EBV receptor, CD21. Even when CD21 is transferred to epithelial cells, EBV can replicate only inefficiently. Latency is then probably established by transfer of virus from the epithelial cell to infiltrating B cells. Thus, circulating anti-EBV IgG may not be as effective as anti-EBV IgA at prevention of the initial steps in infection, whereas anti-EBV CTL may prevent the establishment of latency. It should be noted that there is some experimental evidence to suggest that dimeric IgA may provide a route of infection of epithelial cells that lack the CD21 (CR2) B-cell EBV receptor.

### 9.3.4  Which antigens are likely to produce a useful protective response?

As discussed above, an ideal vaccine would stimulate virus-neutralizing antibody to prevent virus infection and CTL to control the latently infected cell. A virion glycoprotein, gp340/220, has been the target of concerted efforts to generate virus-neutralizing antibody, and the product of EBV nuclear antigen (EBNA) 1, one of the latency genes, would be the ideal antigen to generate CTL. However, anti-EBNA1 CTL have, so far, not been demonstrated (see Section 9.4).

### 9.3.5  Which vaccination strategy is the most appropriate to deliver the relevant antigens?

Traditional approaches to the development of an EBV vaccine are unlikely to succeed primarily because there is no fully permissive cell-culture system, making it difficult to obtain this virus in large enough quantities. Although nontransforming strains of EBV have been described and proposed as live vaccines, even if the problem of virus growth could be overcome, there are insufficient markers of attenuation or appropriate animal models to verify the safety of these viruses. Low yields of EBV also make a killed whole-virus vaccine impractical. Thus, a molecularly engineered vaccine is the only practical alternative. Subunit vaccines are the method of choice for most vaccine manufacturers, and hence the production of gp340/220 (see Section 9.4) for generating EBV antibody has dominated the search for an EBV vaccine. More recently, the identification of CTL epitopes recognized by specific HLA haplotypes means that peptide vaccines can become a reality for generating anti-EBV CTL (see Section 9.4). Other approaches to generating EBV CTL, for example live vectors based on some of the latent proteins, may also prove fruitful.

## 9.4 Identification and characterization of EBV antigens

Cell lines derived from BL biopsies or by tranformation *in vitro* of primary B cells with EBV are the standard source of virus and virus-encoded proteins. Such cell lines can be induced to produce small amounts of EBV and virus proteins by the addition of anti-Ig antibodies or low concentrations of sodium butyrate or phorbol esters. Much of the initial characterization of anti-EBV antibody has been studied using these lines.

Attempts to generate an EBV vaccine have focused on the membrane antigen (MA) originally defined serologically as being on the surface of EBV-containing B-cell lines. MA was subsequently shown to be composed predominantly of the glycoproteins gp340, gp220 and gp85, as well as the capsid protein p140. In the late 1970s and early 1980s, membranes from producer cell lines such as P3HR-1 and B95-8 were shown to raise EBV-neutralizing antibody. Subsequently, gp340 and gp220 were shown to be structurally related, being derived from a single gene by an in-frame splice, and to be the predominant glycoproteins of the virus particle. They play an important role in virus entry and penetration into the B cell by interaction with CD21, the C3d complement receptor (see *Figure 9.5*).

Anti-gp340/220 monoclonal antibodies will prevent EBV infection and abolish its ability to transform B cells. Antibodies raised against purified gp340/220 will neutralize virus and the majority of virus-neutralizing antibody in the sera of EBV-infected individuals is directed against gp340/220. More importantly, it has been shown that immunization with purified gp340/220 can prevent EBV-induced lymphoma in cottontop tamarins (see *Table 9.6*). Consequently, most efforts to generate an EBV vaccine have centered on gp340/220 and its use to induce high levels of virus neutralizing antibody.

### 9.4.1 DNA sequence analysis of gp340/220

Several interesting features of the gp340/220 molecule were elucidated when the DNA sequence became available. The predicted molecular weight of the gene product based on the amino acid sequence is less than 100 kDa, but the actual molecular weight as determined by physical

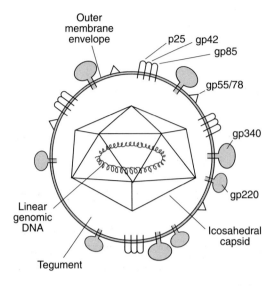

**Figure 9.5:** Diagrammatic representation of the structure of EBV. gp340/220 interacts with CD21 (CR2) and facilitates the entry of the virus into cells bearing CD21. gp85, gp42 and p25 are found as a complex. There is also some evidence that gp55/78 is on the surface of virus particles.

methods is 340 kDa. The difference is due to extensive glycosylation of the molecule. There are 38 potential N-linked glycosylation sites plus 163 threonines and 103 serines as potential O-linked glycosylation sites. The secondary structure is predicted to be a typical type-1 membrane glycoprotein with an amino-terminal signal sequence, hydrophobic membrane anchor and carboxy-terminal cytoplasmic tail. The sequence also reveals splice donor and acceptor sites that would generate mRNA capable of coding for gp220. The region of gp340 not present in gp220 contains a number of imperfect repeats, the exact number of which depends on the strain of virus analyzed. Other interesting features revealed by the sequence are the high serine and threonine content of the molecule (17.5% and 12.5% respectively) as well as a high proline content (12%).

*Figure 9.6* shows a compilation of the antigenicity index predicted by the algorithms in the University of Wisconsin GCG sequence analysis package. Glycosylation and other modifications, such as the addition of sialic acid, complicate the prediction of hydrophilicity and antigenicity of the molecule. Pictorial representations of predicted secondary structures can be seen in the squiggles plot shown in *Figure 9.7*.

### 9.4.2 Characterization of gp340

It is widely recognized that computer predictions of epitopes are at best unreliable and often misleading. It is necessary, therefore, to determine

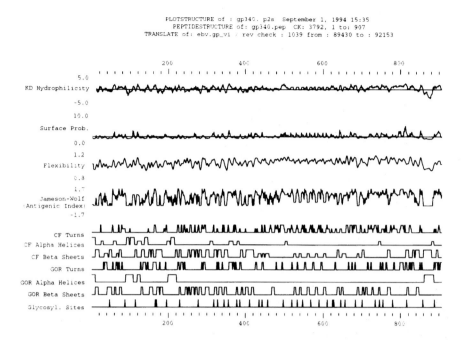

**Figure 9.6:** Compilation of primary sequence features and antigenicity index for EBV gp340 as predicted by the GCG suite of programs. Printout from the Wisconsin sequence analysis package reproduced with permission from Genetics Computer Group Inc.

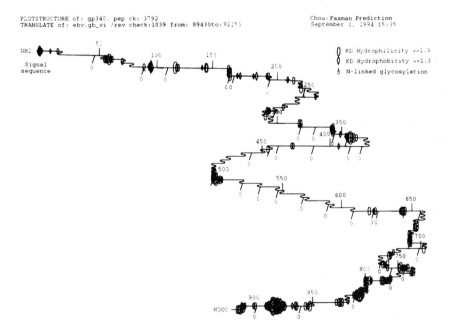

**Figure 9.7:** 'Squiggles' secondary structure plot of EBV gp340 from the GCG analysis programs. Printout from the Wisconsin sequence analysis package reproduced with permission from Genetics Computer Group Inc.

experimentally the antigenic regions of molecules. This can be done in a variety of ways. Fragments of gp340 were expressed in *E. coli* or in eukaryotic cell lines and analyzed for reactivity with anti-gp340 monoclonal antibodies raised against the authentic product. *Table 9.2* shows the reactivity of various monoclonal antibodies with *E. coli*-expressed fragments of gp340 and two preparations of gp340 derived from eukaryotic cell lines. Very few of the antibodies react with the bacterial products whereas most react with the native molecule, confirming that linear epitopes, as represented by the bacterial proteins, are less important than discontinuous epitopes (conformational epitopes) in neutralization of EBV by antibodies. In prokaryotes, recombinant eukaryotic cell proteins are not glycosylated or post-translationally modified in the same way as EBV-derived gp340. Indeed, a protein expressed in *E. coli* representing over two-thirds of gp340 produced only weak virus-neutralizing antibody after 12 immunizations of rabbits with 500 μg of protein!

In mice, epitope group I seems to be an immunodominant epitope with a significant percentage of monoclonal antibodies being directed to this epitope. Interestingly, it is also a virus neutralizing epitope. Using fragments of gp340 expressed in eukaryotic cells, it has been shown that the amino-terminal region of the molecule (460 amino acids, but possibly also a much smaller fragment of 162 amino acids) reacts with epitope

**Table 9.2:** Reactivity of purified gp340/220 with various monoclonal antibodies

| Epitope group | Antibody | BPV | Reactivity with gp340 fragments expressed in *E. coli* | B95-8 |
|---|---|---|---|---|
| I | 72A1[a] | + | − | + |
|  | F30 3C2 | + | − | + |
|  | F34 4E3 | + | − | + |
|  | F29 1G7[a] | + | − | + |
|  | F34 1F2[a] | + | − | + |
|  | F34 5D3 | + | − | + |
|  | F34 6B1[a] | + | − | − |
|  | F34 6B5 | + | − | + |
| II | 2F5 6 | + | − | − |
| III | F16 3E3 | +/− | − | − |
| IV | F29 89[a,b] | +/− | 733-841 (FP6,7) | + |
| V | F16 11B7 | + | − | + |
|  | F16 1C10 | + | − | + |
| VI | F34 5H7 | + | 326–429 (FP4) | + |
|  | F34 2B11 | + | 326–429 (FP4) | + |
| VII | F30 5C8 | +/− | − | − |
|  | F34 1D8 | − | − | − |

Reactivity of monoclonal antibodies from seven epitope groups with the purified BPV-derived gp340/220 was determined by ELISA. Data in the column headed B95-8 show reactivity in an immunoprecipitation assay. '+' indicates a significant reaction, '−' indicates no reaction and '+/−' indicates a weak reaction.
[a]A virus neutralizing antibody.
[b]Although reactivity was low as determined by ELISA, Western blots gave strong signals at high dilution of the antibody.

group I monoclonals. Yet sera from EBV-infected individuals only occasionally react with gp340 amino-terminal *E. coli*-derived fusion proteins, FP2 and FP3. This illustrates the difficulties in interpreting antibody reactivity data using different expression systems and different 'immunized' species.

Synthetic peptides can also be used to map linear epitopes within a molecule (see Section 6.3.5). It is clear from *Figure 9.8* and *Table 9.2* that the epitopes of gp340 recognized by vaccination of rabbits with a variety of immunogens are substantially different from those epitopes recognized after natural infection. A further fact that is clear from the data is that different forms of the immunogen give different specificities of response. Interestingly, with all but two of these rabbits their sera bound strongly to epitopes within the region 236–327 (*Figure 9.8*), while sera from naturally infected humans reacted only weakly with this region. The major

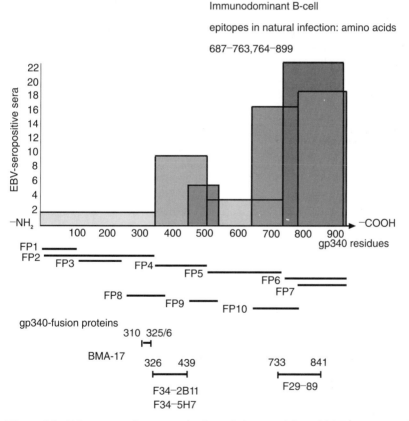

**Figure 9.8:** Diagrammatic representation of the reactivity of 22 EBV-seropositive individuals with gp340 β-galactosidase fusions expressed in *E. coli*. The results indicate the presence of commonly recognized B-cell epitopes in the carboxyl-terminal one-third of the gp340 polypeptide. For details of the fusion proteins see ref. [2]. Reactivity of four monoclonal antibodies is also indicated. Reproduced from ref. [2] with permission from the American Society for Microbiology.

reactivity in the human sera was toward the carboxyl terminus of gp340 and could not be mapped with peptides, indicating that sera reacted with discontinuous epitopes, that is conformationally sensitive epitopes. Interestingly, these antibodies were not virus neutralizing but may be important in ADCC or complement fixation. In neither case was there significant antibody activity detected in the region of gp340 that interacts with CD21 (the amino terminus). Even within one species (rabbit) different forms of the antigen gave a different spectrum of peptide reactivity even though one peptide was recognized by all the antisera.

In rabbits, FP2, 3 and 4 all raised antibodies which bound strongly to purified gp340 but none of the FPs raised antibodies which neutralized virus infectivity (*Table 9.3*). Although these regions may not be important for virus neutralization, they could still be important for virus clearance via ADCC or complement fixation. Antibody to whole virus and to purified gp340 will neutralize viruses and, as indicated by the monoclonal antibody data, conformational (discontinuous) epitopes are crucial in virus neutralization. The implication of these data for vaccine production is that bacterial expression systems and other approaches that rely on linear epitopes are not likely to be successful in the production of neutralizing antibody.

**Table 9.3:** Summary of peptide-defined linear B-cell epitopes in gp340 from various antigens

| Immunogen | Peptide bound (amino acids) | Bacterial protein | Peptide sequence |
|---|---|---|---|
| FP2 | 199–213 | FP2, FP3 | 199–213 KTEMLGNEIDIECIM |
| | 297–311 | | |
| | 301–315 | FP2, FP8 | 301–311 SNIVFSDEIPA |
| FP3 | 142–156 | | |
| | 145–159 | | |
| | 148–162 | FP2, FP3 | 148–156 NPVYLIPET |
| FP4 | 433–447 | | |
| | 437–451 | FP4, FP9 | 437–447 TGFADPNTTTG |
| gp340-ISCOMs | 233–247 | | |
| | 236–249 | FP2 | 236–247 ESHVPSGGILTS |
| | 301–315 | FP2, FP8 | 301–315 SNIVFSDEIPASQDM |
| | 313–327 | | |
| | 317–331 | FP2, FP8 | 317–327 TNTTDITYVGD |
| Whole EBV | 248–262 | | |
| | 251–262 | FP2, FP8 | 251–262 VATPIPGTGYAY |
| | 297–311 | FP2, FP8 | 297–311 YCIQSNIVFSDEIPA |

### 9.4.3 Identification of CTL epitopes for inclusion in an EBV vaccine

EBV readily infects B lymphocytes *in vitro* and induces their transformation into permanently growing lymphoblastoid cell lines (LCL). LCL

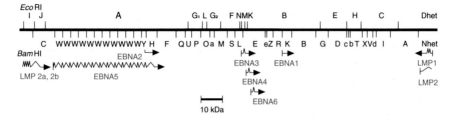

**Figure 9.9:** Physical map of EBV latent genes. The location of EBV latent genes EBNA1–6 and LMP1 and 2 is shown in relation to the *Eco*RI and *Bam*HI restriction fragment map of EBV. Letters indicate the fragment sizes in decreasing order of molecular weight. Arrows indicate the direction of transcription of the latent gene mRNAs and whether they are spliced.

express high levels of activation antigens such as CD23, CD30, CD39 and CD70, as well as adhesion molecules such as ICAM1, LFA1 and LFA3. This suggests that EBV immortalization of B cells may be achieved by the constitutive activation of the same signals that drive B-cell differentiation. Along with the activation and adhesion markers, EBNA1–6 and two latent membrane-bound proteins (LMP1 and LMP2) are also expressed in immortalized lines. *Figure 9.9* shows the physical location of these genes on the background of *Bam*HI and *Eco*RI restriction endonuclease maps of EBV DNA.

EBV-positive B-cell lymphomas arising in immunocompromised patients, such as transplant patients receiving long-term immunosuppressive therapy, AIDS patients and patients with primary immunodeficiency diseases, appear to represent *in vivo* the counterpart of the *in vitro* LCL produced by EBV infection. These tumors express similar cell surface markers to those found on LCL and also express the same range of latent EBV gene products. *Table 9.4* shows the EBV genes expressed in other EBV-associated tumours and thus indicates the genes that might be appropriate targets for the immune control of corresponding tumors and latently infected B cells.

Immune control of tumors in immunosuppressed patients can be exercised through a range of CTL epitopes covering the full spectrum of

**Table 9.4:** Pattern of viral gene expression in EBV-associated malignancies

| Tumor type | | EBV antigens in tumor | % EBV positive |
|---|---|---|---|
| Burkitt's lymphoma | Endemic | EBNA1 | > 95 |
| | Sporadic | EBNA1 | 25 |
| | AIDS | EBNA1 | 40 |
| Immunoblastic lymphoma | PTLD | EBNA1–6, LMP1,2 | 100 |
| | AIDS | EBNA1–6, LMP1,2 | 100 |
| Nasopharyngeal carcinoma | | EBNA1, LMP1,2 | 100 (60%) |
| Hodgkin's disease | | EBNA1, LMP1,2 | > 40 |

**Table 9.5:** Anti-EBV CTL epitopes identified by peptides

| Antigen | Amino acids | Peptide sequence | HLA restriction |
|---------|-------------|------------------|-----------------|
| EBNA2 | 42–51 | DTPLIPLTIF | A2 |
| | 276–290 | PRSPTVFYNIPPMPL | B18 |
| EBNA3 | 331–339 | FLGRAYGL | B8 |
| EBNA4 | 101–114 | NPTQAPVIQLVHATY | A11 |
| | 396–410 | GRPAVFDRKSDAKSY | A11 |
| | 416–424 | IVTDFSVIK | A11 |
| | 481–495 | LPGPQVTAVLLHEES | A11 |
| | 551–564 | DEPASTEPVHDQLL | A11 |
| EBNA6 | 258–266 | RRIYDLIEL | B27 |
| | 281–290 | EENLLDEVRF | B44 |
| | 332–346 | RGIKEHVIQNAFRKA | A24/B44 |
| LMP2 | 236–244 | RRRWRRLTV | B27 |
| | 329–337 | LLWTLVVLL | A2.1 |
| | 426–434 | CLGGLLTMV | A2.1 |

latent EBV gene products. The options for control are more limited in BL and in NPC where antigen expression is limited to EBNA1 and, in some cases of NPC, LMP1 or LMP2.

An heroic amount of work is summarized in *Table 9.5* (see Section 6.4.4 for the methods involved and an example of the prediction and analysis of CTL epitopes in LMP2) which shows the epitopes of EBV latent genes that are presented by the HLA types indicated either to polyclonal CTL or to CTL clones. Several important conclusions can be made from this work:

(1) dominance of certain MHC class I restrictions appears to correlate with preferential recognition of certain viral antigens (e.g. HLA-A2 frequently presents epitopes from LMP2 but HLA-B8 presents epitopes from EBNA3);

(2) in any one individual, EBV-specific CTL responses are frequently a composite of reactivities against different antigens, often with a dominant reactivity;

(3) most CTL responses so far described are to the high molecular weight nuclear antigens EBNA3, 4 and 6;

(4) no CD8$^+$ CTL specific for EBNA1 have been described even though this antigen is expressed in most EBV-positive cells and, in some cases, is the only EBV gene expressed.

LMP1 expression can induce rejection of nonimmunogenic mammary carcinomas in syngeneic hosts whereas EBNA1 expression cannot. This reinforces the idea that this viral antigen may not be recognized as readily or as broadly as other virus antigens. This has led to the speculation that EBNA1 has evolved in such a way as to remove CTL epitopes providing a convenient explanation for the growth of EBV-associated tumors such as NPC in individuals who are not overtly immunosuppressed. However, a class II MHC restricted CTL epitope within EBNA1 has recently been identified which is not processed by B cells but is recognized after the exogenous addition of the peptide epitope.

As latently infected B cells are controlled in healthy individuals by a $CD8^+$ CTL response specific for EBNA2–6, it may be possible to vaccinate against IM by stimulating the appropriate CTL in uninfected individuals. The Queensland Institute for Medical Research is in the process of conducting a vaccine trial to prevent IM, using a peptide vaccine. They have chosen the EBNA3 CTL epitope FLGRAYGL in a water-in-oil adjuvant and are intending to vaccinate EBV-seronegative individuals with the HLA-B8 haplotype. It is hoped this will establish whether such CTL vaccines can protect against IM and post-transplantation lymphomatous disease.

The major difficulty with this approach is that HLA diversity will require either determination of an individual's HLA type before vaccination and then use of an appropriate peptide, or use of a cocktail of CTL epitopes to protect a high proportion of any given population. Another approach would be to engineer a poly-epitopic protein that contained the minimum CTL epitopes required for protection in a large population and to deliver the synthetic protein by vector or with an appropriate adjuvant. The Queensland group constructed a recombinant vaccinia virus expressing just such a poly-epitopic protein which efficiently presented all the EBV CTL epitopes present to their restricting alleles. A further complicating factor can be the necessity for recognition of a peptide by the TCR protein in order for lysis to occur. It has been shown that some individuals, despite an appropriate HLA, do not mount a CTL response against a defined epitope in EBNA3. The majority of other individuals recognize this epitope as a dominant CTL epitope, and it is assumed that the differences in response are due to differences in T-cell receptor proteins.

Another question about the persistence of viruses in the face of a vigorous immune response is being analyzed using EBV as a model. One might expect that pressure to escape from immune surveillance by CTL might result in the loss of defined CTL epitopes by random mutation. This would be more likely to take place where there was a high prevalence of a single haplotype in a population and preference of the HLA type for a single epitope within a protein. Evidence is emerging from sequencing studies of EBV strains that mutation can occur preferentially in some CTL epitopes.

## 9.5 Vaccines based on gp340

Large quantities of gp340/220 will be required for mass vaccination campaigns and, since the genomic location and DNA sequence of the appropriate gene are known, it should be possible to use recombinant DNA technology to provide a source of this glycoprotein. Hence, a variety of expression systems has been employed, including *E. coli*, several

**Table 9.6:** Experimental EBV vaccines in primate systems

| Source of gp340/220 | Animal | Route and number of doses | Neutralizing antibody | Protection | Comments | Year |
|---|---|---|---|---|---|---|
| Protein gel purified from B95-8 cells | S. oedipus | i.p.; 17 | ✓ | ✓ | Large numbers of booster doses required | 1985 |
| Monoclonal antibody purified from B95-8 cells | S. oedipus | i.p.; 6 | ✓ | – | No protection despite high levels of neutralizing antibody | 1986 |
| FPLC purified from B95-8 cells | S. oedipus | s.c.; 5 | ✓ | ✓ | The only derivative of MDP used as an adjuvant | 1989 |
| *Recombinant vaccines* | | | | | | |
| Vero cells | C. jaccus | i.m., 3 sites; 3 | 1/4; 2/4 | 2/4; – | Alum and Freund's as adjuvant | 1989 |
| C-127 cells | S. oedipus | – | ✓ | ✓ | Alum or threonyl derivative of MDP as adjuvant | 1993 |
| Vaccinia virus strain WR | S. oedipus | s.c., $3 \times 2$, $3 \times 1$; $10^7$ p.f.u. | – | ✓ | Protection despite lack of neutralizing antibodies | 1988 |
| Vaccinia virus vaccine strain Wyeth | S. oedipus | s.c., $10^7$ | – | – | | 1988 |
| Vaccinia virus Chinese vaccine strain Tian Tan | H. sapiens | s.c., $5 \times 10^7$ | ✓ | ? | First human trial of an EBV vaccine | 1990 |
| Adenovirus | S. oedipus | – | ✓(weak) | ✓ | Good levels of antibody only weakly neutralizing | 1993 |

live virus vectors, a variety of different eukaryotic cell lines and yeast. gp340/220 undergoes extensive post-translational modification with N- and O-linked glycosyl side chains accounting for over 50% of its molecular weight. It was found that the *E. coli*- and yeast-derived proteins were only poorly immunogenic, possibly due to a lack of glycosylation in the case of the bacterially derived product, or aberrant glycosylation in the case of the yeast-derived product. Thus, the extent of glycosylation is likely to affect the structure and immunogenicity of gp340/220. Because eukaryotic systems are likely to give the product which most resembles authentic gp340/220, all the protection experiments in cottontop tamarins have been carried out with eukaryotic gp340 subunit vaccines or gp340 expressed from eukaryotic viruses. A gp340 vaccine produced from C-127 cells by a BPV vector is now in preproduction development. When vaccine is produced, it is intended to immunize students who are naive for EBV in a phase I/II trial. Prevention of seroconversion to EBV or diminished effects of infectious mononucleosis would be considered as a positive outcome. Trials are unlikely to start for at least a year and would take 2–3 years to complete.

The data in *Table 9.6* indicate, at least in the cottontop tamarin, that protection from EBV-driven lymphoma cannot be correlated with antibody to gp340. In one case there are high levels of neutralizing antibodies but no protection, while in another case there is no detectable antibody yet protection. In the case of the adenovirus recombinants there is antibody but it is mostly nonneutralizing.

Clearly there are significant gaps in our understanding of EBV virus pathogenicity, particularly in the initial infection process, and what constitutes a protective immune response. The planned human vaccine trials using either peptides, to stimulate anti-EBV CTL, or subunit gp340, to produce anti-EBV antibody, will doubtless cast light on these issues. It is likely, however, that both vigorous antibody and CTL responses will be required to protect against EBV-induced disease.

## References

1. Deacon, E.M., Pallesen, G., Niedobitek, G., Cocker, J., Brooks, L., Rickinson, A.B. and Young, L.S. (1993) *J. Exp. Med.,* **177,** 339.
2. Pither, R.J., Zhang, C.X., Shiels, C., Tarlton, J., Finerty, S. and Morgan, A.J. (1992) *J. Virol.,* **66,** 1248.

Chapter 10

# HBV vaccines – from the laboratory to license: a case study

## 10.1 Epidemiology and disease in HBV infection

HBV is a public health problem of global importance. It is estimated that over 300 million people are persistently infected and as such have greatly increased risks of liver damage and primary hepatocellular carcinoma (PHC). PHC alone causes between 250 000 and 1 000 000 deaths per year. HBV-associated pathology varies from a mild acute infection to a persistent carrier status. In high endemicity areas (*Table 10.1*) 7–20% of individuals are chronically infected with HBV, and more than 70% of adults show evidence of prior infection. In these populations almost all infections are acquired in infancy or early childhood and few adults remain susceptible to infection. Infection occurs during the perinatal period (vertical transmission) or during the postnatal period (horizontal transmission) from infected mothers, siblings or other chronically infected individuals. Those who remain susceptible until puberty may become infected by sexual contact.

In intermediate endemicity areas (*Table 10.1*) the prevalence of chronic HBV infection is 2–7% and the overall infection rate is between 20 and 50%. Disease transmission patterns are mixed and disease occurs at all ages, but again the predominant period of transmission is at younger ages.

In low endemicity areas (*Table 10.1*) less than 2% of the population carry HBV and the prevalence of infection is less than 20% in adults. Even though the infection occurs primarily in adulthood, a significant contribution to HBV disease burden is due to transmission at younger ages. Infection rates are relatively uniform for a given area, the widest

**Table 10.1:** Levels of HBV infection and mode of transmission

| Level of infection | Geographical region/ population group | Common mode of infection |
|---|---|---|
| **High** | | |
| 7–20% of population are chronic carriers. More than 70% of adults show evidence of prior infection | Most regions of Asia (except Japan and India) Africa Most of the Middle East The Amazon basin of South America Most Pacific island groups Eskimos Australian aborigines Maoris | Almost all infections are acquired in infancy or early childhood and few adults remain susceptible to infection. Transmission occurs either vertically from chronically infected mothers in the perinatal period or horizontally from infected mothers, siblings, or other chronically infected individuals inside or outside the family. Those who remain susceptible until adolescence may become infected by sexual exposure |
| **Intermediate** | | |
| 2–7% of the population are chronic carriers and 20–50% of adults show evidence of infection | India Part of the Middle East Western Asia Japan Eastern and Southern Europe Most of South and Central America | Disease transmission patterns are mixed and transmission occurs in all age groups. The predominant period of transmission is probably among young children, adolescents and young adults |
| **Low** | | |
| Prevalence of carriage is less than 2% of the population and less than 20% of adults show evidence of infection | United States Canada Western Europe Australia New Zealand | Disease transmission occurs primarily among adults. Transmission in early childhood still occurs |

variations being seen in intermediate endemicity areas where socio-economic factors may play an important role in the relative risk of infection. In low endemicity areas, foci of infection can be seen in some racial groupings such as the Inuits in Canada and the Maoris in New Zealand. Immigration from high or medium endemicity areas can also lead to disparate infection rates in some population groups. However, the general uniformity of risk has greatly simplified the design of immunization strategies to prevent HBV and its sequelae.

The evidence that HBV has a role in the etiology of PHC is compelling and comes primarily from epidemiology. There is a striking geographical correlation between the incidence of HBV carrier status and the prevalence of PHC. Retrospective epidemiological studies have shown that the HBsAg is seen more frequently in the serum of PHC patients than in a control population. The strongest evidence comes from seminal studies by Beasley and colleagues in Taiwan [1]. They showed, in a prospective study of 40–59-year-old males (who have a threefold elevated risk compared to females of developing PHC), that HBsAg carriers were 217 times more likely to develop PHC than a control age-matched population (*Table 10.2*). The data do not exclude a role for other cofactors such as aflatoxin in some areas of China or HCV in Japan but do imply that factors other than HBV chronic infection need not be invoked to explain the epidemiological studies. It is of interest that PHC generally develops after a long history of cirrhosis often of 30–40 years duration. Additional evidence comes from animal studies where woodchuck hepatitis virus, a close relative of HBV, can induce PHC experimentally.

The final proof of a role for HBV in PHC will be if immunization against HBV is shown to decrease incidence of the disease in a high-risk population, but it should be noted that this is a very long-term goal. Immunization of expectant mothers, newborn and infants can be expected to show results only after at least 20 years because the peak incidence of PHC is in middle-aged males.

**Table 10.2:** Relative risk of primary hepatocellular carcinoma (PHC) in individuals chronically infected with HBV (HBsAg + )[a]

| | Cause of death[b] | | | Population at risk[c] | PHC incidence[d] | Relative risk |
|---|---|---|---|---|---|---|
| | PHC | Cirrhosis | Other | | | |
| HBsAg + ve (carrier) | 40 | 17 | 48 | 3414 | 1158 | 223 |
| HBSAg − ve | 1 | 2 | 199 | 19 223 | 5 | 1 |

[a]Data from ref. [1].
[b]In carrier group the biggest single cause of death was PHC; greater than either accidents or heart disease.
[c]Male government workers of Chinese origin in Taiwan. No increased risk was apparent from past HBV infection.
[d]Incidence of death from PHC per 100 000 during 5 year study period.

## 10.2 HBV structure

In the late 1960s it was recognized that sera with high HBV titers contained HBsAg, originally called Australia antigen, and the antibody to this antigen has been widely used in the diagnosis of HBV infection. Large numbers of particles can be seen in the sera of carriers (*Figure 10.1*). These include double-shelled infectious 42 nm virus particles (Dane particles) as well as empty particles approximately 20 nm in diameter, or rods of variable length and 20 nm in diameter (*Figure 10.1*). The outer protein shell of the Dane particle contains the virus surface protein, HBs, also termed HBsAg or env (for envelope glycoprotein) (*Figure 10.2*). HBs is actually a combination of three proteins: small or major, middle and large (S, S + preS1 and S + preS1 + preS2, respectively) which have identical carboxyl termini (*Figure 10.3*). The major protein encoded by the S-gene

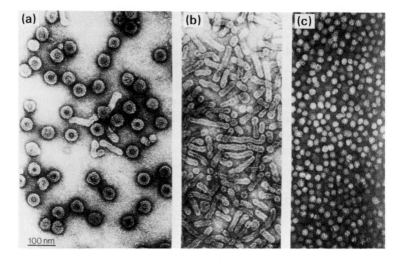

**Figure 10.1:** Particles associated with HBV infection. HBs particles isolated from the plasma of a typical chronic HBV carrier with high viremia were separated by size chromatography and differential centrifugation. The approximate ratios of Dane particle (a) to filaments (b) to 20 nm particles (c) was 1:10:10 000. Reprinted from ref. [2], p. 44, by courtesy of Marcel Dekker Inc.

has 226 amino acids and exists in two forms: an unglycosylated precursor, p24 and glycosylated gp27. The core of the Dane particle is highly immunogenic and composed of the core protein, HBc, also termed HBcAg (*Figure 10.2*). When virions are present in blood, an additional soluble antigen related to the core (capsid) antigen, hepatitis e antigen (HBeAg), can be found. HBs contains a common serological determinant termed a and subtype determinants d, y, w and r. Three main serotypes, adw, ayw and adr, are commonly observed and have different distributions

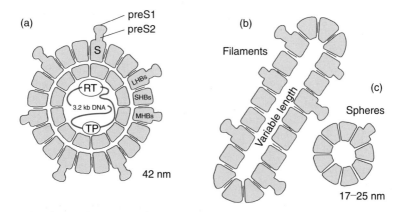

**Figure 10.2:** Model of the HBV particles, filaments and 20 nm spheres. RT, reverse transcriptase; TP, terminal protein. LHBs, MHBs, SHBs; large, medium and small HBsAg. Reprinted from ref. [2], p. 49, by courtesy of Marcel Dekker Inc.

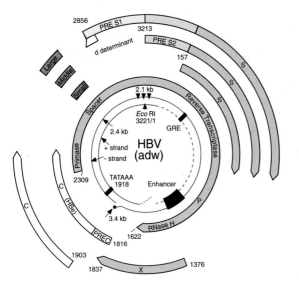

**Figure 10.3:** Genomic organization of HBV (adw2). Nucleotide numbering starts at an *Eco*RI site within the HBV adw strain. The inner circles represent the virion DNA strands and the dashed strand represents the region within which the 3′ end of the + ve strand may occur in different molecules. A line of dots represents the oligoribonucleotide primer covalently attached to the 5′ end of the + ve strand, while the single dot represents the protein primer covalently attached to the 5′ end of the negative DNA strand. The large circular arrows around the outside of the DNA indicate the identified open reading frames, with the direction of transcription from the − ve strand shown. Functions within the polymerase are also noted. HBe antigen is derived from the core region plus the pre-core region (large, middle and small refer to the three 3′-coterminal surface proteins). Small arrows indicate the 5′ ends of three major classes of transcript (3.4, 2.4 and 2.1 kb), all of which are identically terminated near the poly(A) addition signal (TATAA). The mRNAs for the HBe and HBc protein initially run through the stop signal, presenting more than genome length molecules with terminal redundancy. Transcriptional activity is not only governed by promoters but also by an enhancer and a glucocorticoid response element (GRE). C, core antigen; P, polymerase; S, surface antigen; X, X gene.

worldwide. The serotype adw (*Figure 10.3*) arises from an 11 amino acid insertion relative to ayw at the amino terminus of preS1.

HBV has a marked tropism for the liver where HBs interacts with receptors and the virus enters the cell.

## 10.3 HBV vaccines

The structural proteins of HBV, for example HBs and HBc, which are essential in the strategy adopted by a virus for its replication and host-to-host transmission, are also targets for the host's immune system. This may be directed against virus antigens as well as against infected cells and may be mediated by specific antibody to HBs and CTL to HBc.

In general, antibodies that neutralize virus infectivity are directed against components of the virion exposed on the surface of the virus particle. HBV is no different from other viruses in that the antibody to the virus major surface antigen, HBsAg, is sufficient to protect against infection. In 1971 effective immunization of volunteers was shown using heat-inactivated serum from HBV carriers. Subsequent studies showed that the immunogen responsible for such protection corresponded to subviral particles (HBsAg) containing envelope proteins (mainly S-protein). Consequently, the first vaccines to be produced for HBV were based on HBs purified from the blood of HBV carriers.

### 10.3.1 Development phase

*Table 10.3* shows the key steps in production of the plasma-derived vaccine. The first four steps gave material that was about 10% HBs by weight and free of most large virus particles, particularly those of hepatitis B. Further purification and procedures which inactivate any possible contaminant organisms were obviously required. Both of these aims were achieved by pepsin digestion, denaturation with urea and renaturation and formaldehyde inactivation. HBs particles were surprisingly resistant to such treatment and maintained their antigenicity while all infectious agents were rendered inactive. Studies on the immunogenicity of the particles in animals revealed that the potency was enhanced by formulation with an alum adjuvant. The final step in manufacture of the vaccine was co-precipitation from a solution of aluminum hydroxide which gave floccules of the adjuvant in which the antigen was occluded as well as absorbed.

### 10.3.2 Pre-clinical testing

Quality control procedures are very important in the manufacture of vaccines. It is essential to be sure that different batches of vaccine are of

**Table 10.3:** Purification procedure for HBV vaccine purified from the blood of carriers

| Step | | Comment |
|---|---|---|
| 1 | Defibrination of plasma for HBV carriers | Calcium added |
| 2 | Ammonium sulfate precipitation | Concentration step |
| 3 | Isopycnic banding in sodium bromide | Purification step |
| 4 | Rate-zonal separation on sucrose gradient | |
| 5 | Pepsin digestion pH 2 | Inactivation of potentially contaminating infectious agents including HBV |
| 6 | 8 M urea | |
| 7 | Gel filtration | See 3/4 |
| 8 | Formalin 1:4000 for 72 h at 36°C | See 5/6 |
| 9 | Vaccine formulation. 20 µg HBs/dose with 0.5 mg $Al^{3+}$ (alum) in 1.0 ml and thimerisol | |

equal efficacy, purity and safety. In the case of the plasma-derived HBV vaccine, an 'extinction potency assay' was developed where mice were immunized with serial dilutions of the HBs particles. The dilution which gave no immune response to HBs was taken as a measure of the potency of the vaccine. A chimpanzee model was also developed allowing batches of vaccine to be tested for residual HBV as well as providing a model to study protective efficacy of the vaccine. Chimpanzees vaccinated with three doses of vaccine spaced at monthly intervals and challenged with 1000 chimpanzee infectious doses were protected against HBV-associated pathology. Cross-protection also occurred because chimpanzees vaccinated with adw-based vaccine were also protected against challenge with an ayw strain of virus and vice versa. *Table 10.4* shows the plethora of quality control tests required for the HBV plasma-derived vaccine and, although the plasma pool tests are not required for recombinant vaccines, other tests such as DNA content are necessary. General requirements, such as tests for potency, purity, toxicity, pyrogenicity and sterility, apply as much to the HBV vaccine made by recombinant DNA technology as to those derived from human plasma.

### 10.3.3 Clinical trials

At the end of 1975, nine employees of the vaccine manufacturer, Merck, Sharp and Dohme, were vaccinated with a single dose of plasma-derived vaccine and followed clinically for 6 months.

There were no apparent side effects nor evidence of HBV infection, and this opened the way for clinical trials. Initially, they were designed to determine the optimum dose and schedule for the vaccine, and it was found that 75–85% of individuals seroconverted for HBs after two doses, 1 month apart. This was increased to a greater than 95% seroconversion rate

**Table 10.4:** Quality control tests for plasma-derived hepatitis B vaccine

| Plasma pool | Purified bulk antigen | Final tests on formulated vaccine |
|---|---|---|
| Adult mouse (i.p., i.c.) | Microbial sterility | Microbial sterility |
| Suckling mouse (i.p., i.c.) | Blood group substance | Mouse and guinea pig |
| Chick embryo | Human IgM | (general safety) |
| Cell culture | DNA polymerase | Free formaldehyde |
| Vero (monkey kidney) | Gel electrophoresis | Thimerisol |
| WI38 (human diploid) | Protein assay (total and specific | Aluminum |
| | HBs radioimmunoassay) | Identity |
| | Specific absorption | |
| | Formaldehyde | |
| | Pyrogen | |
| | Chimpanzee safety | |
| | Mouse potency (on alum) | |

Reprinted from ref. [4], p. 25 by courtesy of Marcel Dekker Inc.
i.p., intraperitoneal; i.c., intracerebral.

if a booster dose was given 6 months after the initial vaccination. The booster dose is important to increase the responder rates as well as to ensure good immunological memory. Lower seroconversion rates in adults aged over 40 or 50 as well as individuals with immunosuppression have been observed.

The crucial test for any vaccine is a clinical efficacy test. This was first initiated for HBV in 1978 by immunizing male homosexuals in New York City who were at high risk of contracting HBV. Three doses at 0, 1 and 6 months were used, and all vaccinees who responded serologically were protected against HBV infection. Many other studies confirmed the effectiveness of the vaccine, and in 1981 a general license was given to the vaccine. By 1987, 12 commercial concerns manufactured the vaccine worldwide, and over 30 million doses had been administered. The appearance of AIDS threatened the concept of the safety of these plasma-derived HBV vaccines even though HIV was readily inactivated by many of the procedures used in the purification protocol for the vaccine. Consequently, the recent focus for HBV vaccines has centered on the use of recombinant DNA technology to produce HBsAg.

*Table 10.5* shows the effort that has been taken to produce HBs. The comments are intended to show why some of the systems have been followed up, and why others have been abandoned. Currently, the most favored system, particularly for scale-up, is expression of the surface antigen in yeast (*S. cerevisiae*). The production of HBs particles allowed the development of the first licensed recombinant vaccine for human use. Highly purified preparations of such yeast-derived HBsAg particles were shown to be innocuous and to have a high protective efficacy in humans. In 1986 a license was granted in the USA for general use of this recombinant HBV vaccine.

A number of countries have set up trial HBV vaccination projects as a prelude to the introduction of programs for the long-term control of HBV. These trials have often been in remote areas with limited infra-structure in order to ensure widespread vaccination was feasible.

There has been ample discussion on the merits of including hepatitis B immunization in the WHO EPI. The feasibility of this approach was investigated in a trial set up by the Indonesian Ministry of Health in association with the International Task Force on Hepatitis B. The study aimed to immunize one-fifth of the communities on the island of Lombok in 1987 in a manner that was sustainable and applicable to other parts of Indonesia. The Korea Green Cross won the tender to supply vaccine at US¢95 per dose. Prior serum screening confirmed the endemicity of the virus; community education was intensive and staff training thorough, which contributed to the reported high compliance rate. The coverage of children for the third dose of vaccine was 61%, just short of the target 65%. However, 98% of newborns were reached, a remarkable achieve-ment considering the conditions. One direct benefit of the program was to increase the uptake of other EPI vaccines, even in nonproject villages

**Table 10.5:** Expression of the HBsAg in a variety of systems

| Gene/peptide | Vector/host | Year reported/comment |
|---|---|---|
| HBs–β-galactosidase fusion protein | Plasmid/*E. coli* | 1980 |
| HBs–β-lactamase fusion protein | Plasmid/*E. coli* | 1979 |
| HBs | Plasmid/*E. coli* | 1983 |
| HBs peptide in salmonella fimbriae | Plasmid/*S. typhimurium* attenuated strains | 1989. Intended as a prototype for oral immunization against HBV |
| HBs | Plasmid/*S. cerevisiae* (yeast) | 1982. Licensed in 1986 as RECOMBIVAX HB (Merck, Sharp and Dohme) and ENERGEX B (SmithKline Beecham Biologicals) |
| HBs | Chromosomal insertion/*P. pastoris* (yeast) | 1987. Methylotrophic yeast with 5-fold increase in yield over the *S. cerevisiae* system |
| HBs | Chromosomal insertion/*H. polymorpha* (yeast) | 1989. Methylotrophic yeast with 5-fold increase in yield over the *S. cerevisiae* system |
| HBs–HSV gD fusion | Plasmid/*S. cerevisiae* (yeast) | 1985 |
| HBs | Adenovirus/mammalian cells | 1985 |
| HBs | Herpesvirus/mammalian cells | 1985 |
| HBs | Vaccinia virus/mammalian cells | 1983 |
| HBs | Varicella-zoster virus (Oka strain) | 1992 |
| HBs | Alexander cell line (derived from PHC patient) | Investigated by Merck for potential as a source of HBs. Low levels and concerns about use of a tumor cell line in early 1980s |
| HBs | Baculovirus/insect cells | 1987 |
| HBs | SV40/COS cells | 1984 |
| HBs | NIH 3T3 or C127/bovine papillomavirus | 1983. Clinical trial of C127/BPV-derived material (1990) showed no significant difference in safety or immunogenicity from Heptavax B |
| HBs | Mouse LMTK⁻ | 1982 |
| HBs | Vero | 1984 |
| HBs | Rat1 cells | 1982 |
| HBs + preS1 region | CHO cells | 1986. Licensed as GENHEVAC B (Pasteur Merieux et Vacins) |

served by the same vaccinators. Despite the reduced price from the Korea Green Cross, the cost of the vaccine is still a major impediment to the full implementation of the vaccine program.

The government of the Gambia, the International Agency for Research on Cancer (IARC) and the Medical Research Laboratories in the Gambia (UK funded) are jointly responsible for a project funded by the Italian Ministry of Foreign Affairs to incorporate HBV vaccination into the EPI vaccination program. Merck, Sharp and Dohme donated plasma-derived vaccine which, when exhausted, was to be replaced by the yeast recombinant vaccine. HBV vaccine was introduced into the EPI program in 1986 and phased into the whole country by 1990. Overall, 97% of children received one dose, 95% two doses, 91% three doses and 77% received all four doses. One year after vaccination, 724 children were tested, and it was found that 94% were seropositive due to HBV vaccine-induced immunity. Of the remainder, 4% were naturally immune and 0.5% were HBs carriers. Interestingly more than 80% had high levels of anti-HBs ($> 100$ mIU ml$^{-1}$; 10 mIU is considered to be protective).

The indigenous Maori of New Zealand make up about 11% of the population of the country and have a high incidence of HBV infection, as do immigrant Pacific Islanders. In some Maori groups the HBV carrier rate in teenagers is as high as 10% and rates of infection can reach 50% in some communities. Beginning in 1984, all children up to 13 years of age in the Bay of Plenty were offered community-funded vaccination using 2 µg doses of plasma-derived vaccines at 0, 1 and 2 months in hyperendemic areas and at 0, 1 and 6 months in endemic areas. This protocol proved effective with the elimination of clinical infections in vaccinated children. The New Zealand Department of Health now funds universal vaccination of newborn children and routine blood testing of all pregnant women. If a mother-to-be is shown to be a carrier of HBV, the child after delivery will also receive HBV-immune human γ-globulin. In 1988, free immunization was extended to all pre-school (up to 4-years-old) children, and in 1990 a further extension to 15 years of age was approved.

These examples of vaccine production and clinical trials amply illustrate the complexity of converting laboratory-based experimental evidence into a vaccine that prevents disease 'in the field'.

## 10.4 Problems

### 10.4.1 Organizational difficulties

Much of what is listed under further development can be viewed as 'fine-tuning' but a far greater isssue is one of vaccine utilization. Disease caused by HBV is potentially eradicable. There is no animal reservoir, the vaccine is stable, and it is highly effective in breaking the chain of transmission when given to the newborn of infected mothers. Yet, in the United States there are 300 000 new cases of HBV infections every year. The vaccine, although widely available since 1981, has seen limited use in high-risk individuals and most infections occur outside the defined high-risk groups. Worldwide use of the vaccine has been even lower. There are two major reasons for this lack of uptake: universal vaccination, even with a unit cost as low as US$10 per course, would mean an annual bill for vaccinating newborn children in the USA of more than $30 000 000 just for the costs of the vaccine itself. The second reason is that it is difficult, particularly for developing countries, to find funding and to establish vaccine delivery and monitoring systems. Costs for delivery are usually far greater than the cost of the vaccine alone, possibly as much as four times greater! It is clear that vaccination policy will have to be reviewed but, in the light of attempts in the USA to cut the cost of health care, it seems that any change in policy is unlikely to include universal vaccination.

### 10.4.2 Duration of immunity

The duration of protection conferred by the HBV vaccine has been the subject of intense debate for several years. It is clear that after vaccination there is steady decline in the levels of circulating antibody. After 5 years, 90% of vaccinees continue to retain detectable levels of anti-HBs and approximately 80% have levels considered protective (10 mIU ml$^{-1}$). Even loss of detectable antibodies may not correlate with loss of protection. Individuals with no detectable anti-HBs antibodies have intact immuno-logical memory because booster vaccination produces an anamnestic increase in this antibody. This suggests that exposure to natural infection would stimulate the same immune response.

The most thorough studies on the duration of protection have been carried out in homosexual males, Alaskan natives and in health care workers. In one cohort of previously immunized homosexual males, HBV infection was detected in a small number of individuals. This was many years after vaccination and all but one case was subclinical. This group of patients was evaluated in the early 1980s before many men were avoiding high-risk sexual behavior and is likely to represent a 'worst case scenario' with individuals continually being exposed to HBV. In contrast, in a cohort of vaccinated health care workers, no cases of clinical or subclinical infection were encountered. Even in the cited cohort of homosexual men, HBs antigenemia was rarely observed: protection from clinical illness, antigenemia and chronic infection was the rule.

Any uncertainty left is due to a gap in the understanding of how the vaccine works. Does the circulating antibody prevent infection by decreasing the amount of virus to a level that is not effective in establishing infection or does the antibody prevent spread of virus after initial infection? Or both? Infection by HBV in previously vaccinated individuals may induce antibodies sufficiently quickly to prevent the spread, but it is unlikely to prevent initial infection. Based on the apparent long-term protection of vaccine recipients and the lack of evidence that a booster vaccination is more effective than natural re-exposure, there is no formal recommendation for booster vaccination except for hemodialysis patients.

### 10.4.3 Generation of escape mutants

Infection with HBV after vaccination may result in the generation of mutants that escape neutralization by vaccine-induced anti-HBs antibodies (see *Figure 10.4*). Evidence for escape has come from liver transplant patients who received HBs-positive livers. Suppression of detectable HBsAg was achieved initially with a high titer human–mouse hybrid monoclonal antibody. After several months of therapy HBsAg was detected that was not recognized by the hybrid monoclonal antibody. The variants that emerged had a single amino acid change in the HBsAg determinant.

An identical substitution has been observed in a vaccinated infant born to an HBV-infected mother. This child became a carrier despite vaccination, presumably because the virus was not completely neutralized by the induced antibody. Long-term follow-up and surveillance are clearly required to establish the clinical significance of such findings. It may be necessary in the long-term to include the mutant epitope to prevent emergence of the escape mutants.

### 10.4.4 Nonresponders

An increasing number of people fail to acquire protective levels of antibody following HBV vaccination. This population consists of both immunocompromised and immunocompetent individuals. It is not surprising that HIV-infected individuals and cancer patients as well as transplant patients receiving immunosuppressive drugs have a variable

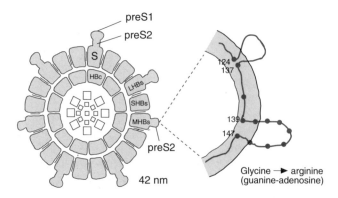

**Figure 10.4:** Proposed structure for escape mutant of HBV. S, surface antigen. (See *Figures 10.2* and *10.3* for detail of preS1 and 2.) Detail derived from ref. [3].

response to vaccination; with from 10 to 70% seroconversion. What is of more concern is that between 2.5 and 5% of immunocompetent individuals fail to make any detectable anti-HBs antibodies following the standard three-dose regimen of vaccination. In these nonresponders, a second course induces a response in about 40% of cases but the response tends to be weak and of limited duration. By comparison, in those people that respond poorly in a first course of vaccination (hyporesponders), but have measurable antibodies, a second course produced an antibody response that is more likely to be adequate and sustained.

Among immunocompetent individuals there is an inverse correlation between the success of vaccination and the age of the person concerned. In young adults there is a seroconversion rate of greater than 95%, but in 60-year-olds the rate is between 50 and 70%. Other factors such as weight, smoking and chronic diseases such as diabetes have not been shown consistently to affect seroconversion rates.

A partial explanation for immunological nonresponsiveness may be particular immune response genes. HLA B8, SC01, DR3 and B44, DR7 and FC31 have been shown to be over-represented in nonresponders and hyporesponders. In a follow-up study, patients known to be homozygous or heterozygous for one of these extended haplotypes were vaccinated; heterozygotes produced antibodies; homozygotes did not. Investigations in mice also point to the fact that responsiveness to HBsAg is an inherited genetic trait. In a murine model using different strains of inbred mice the nonresponsiveness has been overcome by the addition of more epitopes from the preS region of HBsAg. It will be interesting to see the outcome of a recent trial with a yeast-derived HBV vaccine that contains preS2 sequences which are presumably under different histocompatibility control.

## 10.5 Further development

There are a number of incentives for the further development of recombinant HBV vaccines. Higher level expression systems may influence the cost-effectiveness of the vaccine and hence influence immunization schedules. Fewer doses may influence uptake of the vaccine in situations where it is difficult to get people to attend a clinic regularly, if at all. Simplification of administration, for example oral vaccination, would be very valuable for mass vaccination campaigns, particularly by eliminating the need for syringes and needles and requiring less training of individuals who administer the vaccine. Further improvements in efficacy and in the prevention of escape mutants or protection against variant viruses may also be achieved by incorporation of additional protective and T-helper epitopes, hopefully improving responder rates. Some of these improvements will be incorporated into vaccines yet to be fully licensed.

## 10.6 Practical considerations: from laboratory to arm

An enormous amount of work is required to bring a vaccine from the laboratory to the point of being fully licensed. *Table 10.6* illustrates some of the steps in the process and the practical requirements that need to be met in vaccine development. *Table 10.4* shows the array of quality control tests required for the HBV vaccine and illustrates the amount of effort that is required to satisfy regulatory authorities of the quality and safety of biological products for human use.

During the pre-production development stage, plans for phase I safety trials will be drawn up. Full consultation with regulatory authorities for phase II/III efficacy trials will be necessary. If a large enough batch is grown up, this can be used for all trial stages (some should be kept as reference to compare with new batches). Feedback into the development process at all stages of clinical trials increases the flexibility of the process.

### 10.6.1 Timescale

It is clear from the brief review of the development of hepatitis vaccines at the beginning of this chapter that the time taken to move from laboratory studies to widespread use of a vaccine is likely to be more than 5 years.

**Table 10.6:** Requirements of the various stages of vaccine development

| 1. Laboratory phase | Identification and characterization of neutralizing antigen(s), and/or CTL that might be useful in a vaccine |
|---|---|
| | Develop methods for producing antigen–cell lines/recombinant vectors, etc. |
| | Evolve assays for antigen and antibody and CTL |
| | Establish animal model (preferably primate for human disease) |
| | Demonstrate that approach taken for vaccination works experimentally |
| | Timescale 1–5 years |
| 2. Development phase | Freeze down master cell banks, submaster/working cell bank, analyze for presence of adventitious agents (e.g. retrovirus, bacteria, etc.) |
| | Pilot-scale production and purification. Analysis of product, epitopic analysis and quantification |
| | Small animal model of immunogenicity |
| | Optimization of culture and production conditions. Rigorous documentation of all reagents used. Safety and quality control measures are vital in the licensing process |
| | Satisfy regulatory requirements for human use. Formulation (adjuvanting), stability, sterility and immunogenicity tests on final product |
| | Timescale 2–5 years |
| 3. Production phase | Full-scale production, batch process |
| | Storage, adjuvanting, bottling, etc. |
| | Tests on final product (stability, sterility, etc.) |

Following many years' laboratory-based investigations, the first few volunteers to be given a dose of the plasma-derived vaccine were immunized at the end of 1975 and the vaccine was not fully licensed in the USA until 1981. The time taken to license the yeast-derived HBV vaccine was considerably shorter because much was known about the properties of HBsAg particles and experience had already been gained in vaccine trials with the plasma-derived vaccine. However, it still took 5 years for licensing from publication of a paper showing that HBsAg particles could be made in yeast. Part of the reason for this timescale is the length of clinical trials. A fully randomized placebo-controlled phase III efficacy trial will take at least 2 years from initiation to final assessment of the data. If the endpoint is protection against disease, then the timescale may be even longer.

### 10.6.2 Costs

It is difficult to estimate the total costs of vaccine development because experience gained on one project is often useful to another and many different organizations may have played some part in the project. *Figure 10.5* shows estimates for the costs of bringing a new vaccine to the market in the USA (figures from Lederle-Praxis Biologicals). It is worth noting that basic research costs are less than 15% (including all the pre-production development required to optimize production and purification of the vaccine) of the total cost and that clinical trials, together with the cost of borrowing money, are responsible for half the total budget.

Vaccine development is often driven by commercial considerations. UNICEF estimates that the entire global vaccine market is worth $3 billion whereas $3.5 billion was received from sales of the ulcer drug Zantac alone. Due to the high cost of development, the vaccine market is therefore likely to give a poorer return on investment than the pharmaceutical market. Unfortunately, developing countries, with the highest health burden from infectious and parasitic disease and which are most likely to benefit from vaccines, are least able to pay for them.

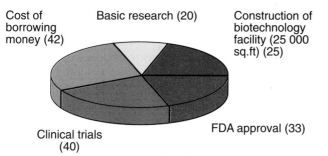

Cost of borrowing money (42)   Basic research (20)   Construction of biotechnology facility (25 000 sq.ft) (25)

Clinical trials (40)   FDA approval (33)

**Figure 10.5:** Average vaccine development cost (figures from Lederle-Praxis Biologicals). It costs approximately US$100 m–US$160 m to develop a new vaccine.

Companies make money if a vaccine is used in a developed country and therefore vaccines for developing countries, particularly to protect against parasitic diseases, receive a lower priority.

This skew to the interests of developed countries, understandable though it may be, is also seen in other areas. The HIV vaccine effort is focused on strains of virus prevalent in the USA and Europe where arguably the problem is less serious than elsewhere. Also, HIV vaccines developed to date – such as those made by Genentech, Chiron, MicroGenSys and Immuno – are all based on genetically engineered versions of gp160, the virus envelope glycoprotein. They are perceived as safer than traditional vaccines such as inactivated or attenuated viruses which might be more appropriate for developing countries. Thus safety criteria and risk assessments of developed countries have dictated the approaches taken in vaccine development. A further consequence of vaccines requiring advanced technology is that developing countries will not be able to manufacture these new generation vaccines.

Costs are not limited to the vaccine itself but as much as four-fifths of the costs of a vaccine program can derive from establishing a developed infrastructure. This often means setting up a cold chain to maintain vaccine stocks at 4°C and training personnel to administer a vaccine. It also involves costs associated with organizing administration, and transporting the vaccine. Developing countries with minimal healthcare budgets may simply not be able to afford to vaccinate their citizens even if very cheap vaccines were available.

It should also be noted that the costs, return on investment, timescale of vaccine development and potential litigation from side effects of vaccination have all discouraged companies from producing vaccines. The consequence of this is that there is only one vaccine manufacturer left in the UK and only a few companies actively involved in vaccine research. The situation is more positive in the USA, where there are at least five well-financed companies involved in vaccine development and production. There has also been a proliferation of small companies, founded with venture capital, whose expertise is in a single new area of technology (e.g. companies developing BCG vectors or vaccinia vectors or adjuvants).

There is a complex mix of interacting interests in vaccine development and use. UNICEF purchases vaccines from 12 manufacturers in Europe and Japan at discount prices and supplies approximately 40% of all vaccines in the developing world. Not one US firm bid on UNICEF tenders, at least in part because when some firms indicated their intention they were denounced at US congressional hearings for charging high prices to the United States while developing countries got much lower prices. Vaccine companies in the US are also complaining about legislation that fixes the price of vaccines for use in President Clinton's children's vaccine initiative. They claim, with some justification, that lower profit margins will only reduce spending on research and development.

### 10.6.3 The politics of vaccination

Immunization is a complex and at times fragile enterprise requiring the involvement of large numbers of organizations. In the USA it has been estimated that 20 federal agencies, as well as state departments of health, vaccine and biotech companies, professional medical societies, voluntary organizations and many health care workers are all involved in the production and delivery of vaccines. This elaborate network of interested parties is at best fragmented, with a variety of aims and objectives. The uncoordinated nature of vaccine development is evident from an analysis of what drives progress, which seems to owe as much to market forces as it does to clearly defined public need. In the Philippines, Vietnam and Tanzania, over 80% of children are immunized by the age of 2 years against six childhood diseases. In the USA, a CDC report suggested that less than 50% of children in the cities they surveyed were similarly immunized. President Clinton's childhood immunization initiative aims to increase that figure to 90% by 1996. Coverage in the UK for measles, polio and DPT has already surpassed the 90% figure.

A variety of nongovernmental organizations, including WHO, UNICEF, the United Nations, various foundations (e.g. Rockefeller and Wellcome), as well as various governmental aid programs, also provide support and some direction to the vaccine effort, particularly in efforts to develop vaccines and vaccine programs for developing countries. The Children's Vaccine Initiative (CVI) was launched in 1990 at the World Summit for Children in New York. UNICEF, the United Nations development program, the Rockefeller Foundation, the World Bank and the WHO are sponsoring the initiative which aims to catalyze research and development for new and improved vaccines, with the ultimate aim of producing a 'supervaccine'. These agencies and many vaccine researchers recognized that there were fundamental problems in reaching the world's children with the currently available vaccines, let alone any more that might be developed. To provide protection against measles, diphtheria, tetanus, polio, pertussis and tuberculosis, health care workers would need to visit a child six times in the first 2 years of its life, which is practically impossible in developing countries. To make matters worse, many vaccines require refrigeration and are injected, both of which are difficult to achieve in poor communities. The aim of the CVI is to develop vaccines that are available to all people, require one or two rather than multiple doses, do not require refrigeration, can be given earlier in life and in combinations to reduce the number of visits. There was also a commitment to develop new vaccines to protect against diseases not covered by any current vaccine. It has been estimated that $300 million will be required by the year 2000 if the program is to be successful. As yet only $10 million has been raised and hopes for more money will depend on strong leadership and the political will of more governments such as the USA and UK who as yet have shown little interest.

Even if the optimism of new insights in immunology and molecular biology is justified and yields new approaches, it is likely that political, economic and social considerations will play a large part in the future of vaccines and vaccination.

## References

1. Beasley, R.P., Hwang, L.-Y., Lin, C.-C. and Chien, C.S. (1981) *Lancet,* **ii,** 1129.
2. Gerlich, W.H. and Bruss, V. (1993) in *Hepatitis B Vaccines in Clinical Practice* (R.W. Ellis, ed.). Marcel Dekker, New York, p. 41.
3. Carman, W.F., Zanetti, A.R., Karayiannis, P., Waters, J., Manzillo, G., Tanzi, A., Suchkerman, J. and Thomas, H.C. (1990) *Lancet,* **335,** 325.
4. Hilleman, M.R. (1993) in *Hepatitis B Vaccines in Clinical Practice* (R.W. Ellis, ed.). Marcel Dekker, New York, p. 17.

Chapter 11

# Discussion

## 11.1 Why vaccinate? – prevention is better than cure

The control of infectious and parasitic diseases is one of the most striking achievements of medical science during the 20th century. As described in earlier chapters, this has been accomplished in many instances by the use of vaccines. But can communicable diseases be controlled by other means? To be most effective, any alternative measures must share with vaccines the necessary properties of selectivity and specificity.

A short time after Louis Pasteur demonstrated how vaccines can give protection (immunoprophylaxis) against communicable diseases, Paul Ehrlich (1854–1915) was developing an alternative approach to their control by treatment with specific chemicals (chemotherapy). Ehrlich's idea of drugs that can act against the infecting organisms without damaging the host was conceived following his discovery in the laboratory that certain dyes killed trypanosomes. He subsequently developed the concept of 'magic bullets' – drugs with selective toxicity against specific pathogens. Ehrlich did, in fact, discover one of the earliest chemotherapeutic agents, an arsenical (Salvarsan), which was used in the treatment of syphilis. However, such ideas met with considerable resistance, and chemotherapy did not really start to enter the modern era until 1929 when Alexander Fleming discovered the first antibiotic, penicillin, although this was not used therapeutically until more than a decade later. Antibiotics are natural products of microbial origin that act only against prokaryotic metabolic pathways found in bacteria. This provides the basis for their selective action against bacterial infections in a human host, but it also means antibiotics are virtually ineffective against other infectious agents. Because they share certain properties with bacteria, chlamydial infections may also be treated with antibiotics, although they must be taken up by the host cell in order to be effective. Consequently, extended therapy is essential for total elimination of these organisms from an infected host. Other drugs have become available for the treatment of viral, protozoal and other parasitic diseases but they have certain, sometimes serious, deficiencies.

The attempted control of communicable diseases by chemotherapeutic agents has a number of unavoidable limitations. A common problem arises from the interval after infection before overt clinical signs of disease become obvious. This sub-clinical incubation period can last from a few days up to several weeks or months following infection. Only after characteristic clinical signs or symptoms of disease have appeared does chemotherapeutic treatment usually begin. But by this stage damage has already been caused to tissues and/or organs of the infected individual, and these pathological changes may be well advanced, possibly irreversible or even lethal. Also, the infectious load will have increased markedly during the incubation period so that, by the time an infection does become clinically recognizable as a disease, the pathogen may have reached such numbers that effective therapy has become more difficult. An obvious remedy would be to use drugs prophylactically, that is to prevent infection. However, very stringent safety requirements are expected of any drugs that are administered to apparently healthy people. Further problems arise with recurrent infections and chronic diseases which require either repeated or extensive courses of drug treatment. Prolonged administration can lead to an accumulative toxic effect of the therapeutic agent itself, and this has proved to be a particular drawback of drugs used to treat parasitic diseases.

However, a more crucial limitation of chemotherapy is the development of drug resistance. This phenomenon was first recorded nearly a century ago with protozoal parasites, but it has now become a very serious problem with drug-resistant bacterial infections. Some bacteria, including *S. aureus* and *M. tuberculosis*, have even become multiply resistant to different antibiotics. One answer to drug-resistant bacterial infections has been combination therapy using two or more drugs simultaneously. However, this is not always possible because the therapy depends on the availability of various drugs that are not cross-resistant.

Chemotherapeutic treatment of viral diseases is a relatively recent innovation. One of the first antiviral drugs to be used on a relatively large scale was metisazone (Marboran, $N$-methyl isatin-$\beta$-thiosemicarbazone). This drug was evaluated in clinical trials for the treatment and prevention of smallpox. Although there was little effect on the case:fatality ratio when used therapeutically, metisazone did have some prophylactic effect. In Madras in 1962 it was shown that treatment of 1100 subjects with metisazone resulted in three cases of smallpox but no deaths whereas, in a control group of 1126 untreated subjects, there were 78 smallpox cases and 12 died. However, there were toxic side effects of the drug with many cases of severe vomiting and nausea which in different clinical trials ranged from 25% to 70% or more of treated subjects. Vaccination remained the method of choice for smallpox control and its eventual eradication.

The further development of antiviral chemotherapy has proved particularly difficult. The major problem is the dependence of virus replication on host metabolic pathways. This makes any selective target difficult to identify unless there are unique viral enzymes in the infected cells. In 1977 the discovery of a true antiviral drug, acyclovir (Zovirax, 9-(2-hydroxyethoxymethyl)guanine), was reported. Acyclovir is a potent inhibitor of HSV. It is activated by a virus-coded enzyme, thymidine kinase, to act against a second viral enzyme, DNA polymerase. However, acyclovir and other anti-herpes drugs are ineffective against latent infections. Zidovudine or AZT (3′-azido-3′-deoxythymidine) is another antiviral drug of fame (and infamy!) that inhibits a viral enzyme, the reverse transcriptase of HIV. The anti-HIV properties of AZT were discovered in 1985, and the results of clinical trials were reported in 1987, the same year that the toxicity of AZT in AIDS patients was described. Two years later, in 1989, HIV isolates from AIDS patients on AZT treatment were found to have reduced sensitivity to the drug. It is now apparent that such problems as toxicity and resistance are common deficiencies of the first generation of anti-HIV drugs. Experience of the prolonged administration of other antiviral agents was limited until relatively recently but acyclovir-resistant HSV infections are now seen with increasing frequency in AIDS patients.

Of the anti-parasitic drugs, chemotherapy of malaria has a long history, particularly in the context of medicinal plants. Quinine, the first drug used to treat malaria, was obtained from the bark of *Cinchona* trees. Although used for treatment, quinine is potentially toxic and unsuitable for prophylaxis. Chloroquine is an effective antimalarial for both prevention and treatment, but resistant strains of *P. falciparum* have emerged, particularly in southeast Asia, South America and Africa. Pentamidine and suramin are drugs used to treat the systemic stage of trypanosomal infections but toxic arsenicals are required once the parasite has entered the brain. The few drugs – antimonials, pentamidine and amphotericin B – that are available to treat the early stages of Chagas' disease are toxic and ineffective against the late complications. Onchocerciasis has been treated with diethylcarbamazine and suramin but these drugs may cause severe adverse reactions, including death, although a new drug, ivermectin, is now available. However, praziquantel, a drug considered to be the most effective treatment for schistosomiasis, has recently been found to produce a low cure rate and an unusually high number of severe side effects during its use in an outbreak in a community on the Senegal river in West Africa.

Antibiotics and chemotherapeutic agents clearly have a place in the control of infectious and parasitic diseases (in fact, the recent plague epidemic in India was controlled by the use of streptomycin or tetracycline), but they have limited value because of their toxicity and the emergence of drug-resistant organisms. Also, many antimicrobial drugs are very expensive, particularly when they are used in combination therapy. The need for repeated or continuous treatment adds significantly to the financial costs which are often beyond the health care budgets of developing countries.

## 11.2 Herd immunity and its control of infectious and parasitic diseases

Although the immediate aim of both chemotherapy and immunization is to protect the individual, the next objective must be to protect the community. This may best be done with vaccines. Unlike drug treatment, the induced immune response is long-lasting, and it is not necessary to vaccinate the whole population.

Some infectious agents such as viruses, bacteria and certain protozoa multiply directly within their respective hosts at a rate that is usually rapid in relation to the lifespan of the host. In the case of an acute disease, recovery is due to the host's immune responses which confer durable resistance to further infection. Therefore, if the pathogen is to survive, it must find new hosts, and a pathogen's further survival in a human population is determined by the presence of susceptible individuals. This will decrease, however, as the disease spreads because the proportion of resistant, immune individuals will increase. Eventually 'herd immunity' develops to a level where acute infectious diseases will die out unless susceptible individuals continue to appear within the population. Newborn children can clearly supply this requirement and, if the birth rate is sufficient, a balance can be reached when the disease becomes endemic. The population size that is needed to maintain an endemic disease can vary considerably. In the case of an acute infection such as measles, a minimum population size of about 500 000 is required. However, another viral infection, chickenpox, can remain endemic in much smaller communities of fewer than 1000 individuals. This may be explained by the failure of infected hosts to eliminate the VZV after the first clinical episode. It remains latent in a noninfectious state to be re-activated, perhaps several decades later, to cause shingles. This provides an opportunity for primary infection of those susceptible individuals who have appeared in the population between the two episodes.

Because some endemic diseases are maintained by the rate at which new susceptibles are introduced into the population, such viral or bacterial infections may be controlled by immunization during early childhood. A newborn child will be protected by maternal antibodies for several months, but after that time there is an increasing risk of infection which varies between different geographical locations. The average age at infection with measles is 1–2 years in a developing country, compared with 4–5 years in developed countries before immunization was introduced. There is clearly a window of opportunity for childhood immunization and this is reflected in the chosen vaccine administration schedules (*Table 11.1*).

**Table 11.1:** Immunization schedules in England and Wales

| Vaccine | Age of administration |
| --- | --- |
| Diphtheria, tetanus, pertussis (DTP) | 2–6 months; booster at 4 years |
| Measles, mumps, rubella (MMR) | 12–18 months; rubella only at 14 years to seronegative girls |
| Polio (Sabin) | Same time as DTP; booster at 4 and 16 years |
| BCG (tuberculosis) | 14 years |

A sufficient frequency of immune individuals or degree of 'herd immunity' can be attained artificially by extensive vaccination that makes it impossible for an endemic disease to spread within a population. Immunization coverage required to block transmission of poliomyelitis is 80–85%, but 92–95% in the case of measles.

Although individuals may be protected, mass immunization will not have the impact on parasitic diseases that has been achieved with other infectious diseases. Their occurrence and spread is rather different because there is no direct link between infected individuals and population size, as seen with viral and bacterial infections. Most parasitic protozoa and helminths must leave the human host to find another host of a different species before their life cycle can continue. This transmission is often effected by water or by insect vectors. Vast amounts of money and human resources have been wasted on various unsuccessful attempts to control such parasitic diseases by elimination of the vector. For example, the use of the insecticide DDT to eradicate the mosquito vector of malaria was eventually abandoned when DDT-resistant mosquitos emerged and the toxic effects of DDT on other animals were recognized.

## 11.3 Elimination of an infectious disease: smallpox eradication

Vaccination as a means to control an infectious disease on a global basis began in 1959 when the WHO decided to undertake the eradication of smallpox. Even in 1967 there were an estimated 10–15 million cases each year with 1.5–2 million deaths. In addition to humanitarian considerations, benefit could also be expressed in economic terms. The costs of care given to smallpox victims and the resultant loss of productivity in those countries where the disease was endemic, together with the costs of protection against importation of smallpox into those countries that were free of smallpox, were calculated to be US$1350 million. Approximately 977 million people then lived in the endemic areas and the cost of total eradication, at US$0.10 per person, was initially calculated to be US$97.7 million. Although the final total cost was about US$400 million, smallpox eradication was calculated to save the USA the total of all its contributions to the program every 26 days.

A number of factors were pertinent to the successful eradication of smallpox. Human infection was almost invariably followed by disease, with clinical signs making the disease readily recognizable wherever it appeared. Because smallpox virus is species specific, no animal reservoirs of the disease existed. The vaccine was different to the original cowpox virus used by Edward Jenner, being based on vaccinia virus which, although it belongs to the same genus, has uncertain origins. In the field, vaccine was available as a heat-stable, freeze-dried preparation that could be immediately reconstituted by resuspension in sterile distilled water. It was administered intradermally using a cheap device that could be disposed of after each vaccination. This was a bifurcated needle with sharpened tips that were separated sufficiently to retain by capillarity a volume adequate for one inoculation when the needle was dipped into the liquid vaccine. Initially, mass vaccination was attempted but later in the campaign this policy was changed to the tracking down of smallpox cases and vaccinating all contacts. The last known case of smallpox was reported from Somalia in 1977, and successful global eradication of the disease was officially declared 2 years later on 9 December 1979.

## 11.4 The argument for expanded programs of immunization

The arguments outlined above that led to the eradication of smallpox may also be applied to other infectious diseases. The incidence of several communicable diseases that are the most common enemies of children in developing countries could be reduced substantially by better standards of public hygiene and improved water supplies. It is estimated to cost US$1260 per life saved in rural areas and about US$20 000 per life saved in urban areas to provide adequate sanitation and drinking water. However, six vaccinations cost approximately US$10 per child. Vaccination programs in developing countries have, in fact, markedly reduced the incidence of several childhood diseases. Deaths of under 5-year-olds from measles have been reduced from 2.5 million in 1980 to just over 1 million in 1992; during the same period nonfatal cases of this major cause of disability and malnutrition in children have fallen from approximately 75 million to 25 million per year. Similar successes have been accomplished with whooping cough and neonatal tetanus (*Figure 11.1*). In 1994 the Pan-American health authorities announced the eradication of poliomyelitis from North, Central and South America, and this disease should be eradicated from most other countries of the world by 1995. These achievements are due to increased immunization coverage which has reached 70% or more of 1-year-olds in most of the developing world (*Figure 11.2*). Economic benefits also accrue from vaccination programs: in addition to decreased demands on medical and midwifery services, reduced perinatal and infant mortality leads to a reduced birth rate and the prevention of disabilities increases the adult workforce.

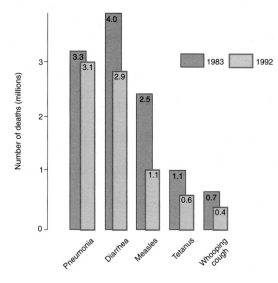

**Figure 11.1:** Deaths of under 5-year-olds from major diseases of childhood in the developing world.

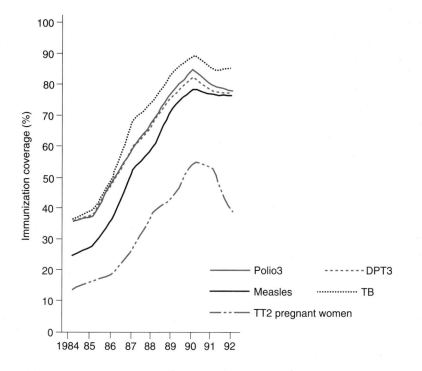

**Figure 11.2:** Immunization coverage. Percentage of the developing world's 1-year-olds protected against the major vaccine-preventable diseases.

## 11.5 The need for new or improved vaccines

Respiratory tract infections are caused by many etiological agents, particularly viruses and bacteria. A very common bacterial cause of acute, community-acquired pneumonia is *S. pneumoniae*. Even in developed countries this pathogen can be responsible for nearly 70% of all cases of lobar pneumonia. Vaccine development is hindered by the antigenic diversity of *S. pneumoniae* which has 84 serotypes (see Chapter 5). However, vaccines based on a cocktail of selected serotypes are available. Recently, it has been recommended that a vaccine combining the most common (nine) serotypes would be adequate for use in developed and in developing countries. Trials will be carried out in children of diverse racial origin, HIV-seropositive children and children with parasitological evidence of malaria infection. As with the *H. influenzae* type b (Hib) vaccine, the pneumococcal vaccine will be based on capsular polysaccharide conjugated to carrier protein. Clinical trials of acellular pertussis vaccines have also given promising results. Such subunit vaccines are based on pertussis toxin; the filamentous hemagglutinin which is involved in attachment of the organism to mammalian cells; and fimbrial proteins that are potential adhesins. Further development of these bacterial subunit vaccines could clearly be based on other strategies which employ recombinant DNA or anti-Id technologies.

*M. tuberculosis* is responsible for 10 million new cases and 3 million deaths each year on a global basis. Until recently such statistics applied mainly to the developing countries but between 1986 and 1992 there were nearly 52 000 excess cases of tuberculosis (TB) in the USA. This figure was calculated by comparison with the numbers expected if the previous decrease in TB cases between 1981 and 1984 had continued. The increased incidence of the disease in the USA has been accompanied by a rise in multiple drug-resistant TB. Elsewhere in the world there is an emergence of TB resistant to isoniazid and rifampicin. To this sad catalog may be added the dismal prediction that the global incidence of TB will increase further due to the recent appearance of HIV and the spread of AIDS in both developed and developing counties.

Although the BCG vaccine has now been available for over 70 years, there are still doubts concerning its suitability. This is due mainly to the considerable variation in its efficacy of protection ranging in different trials from 0% to over 70% following BCG vaccination (*Figure 11.2*; see also *Table 11.2*). Failures were given various explanations, the most popular being differences between BCG strains. Rather than a single seed strain, BCG has been manufactured in many different countries from their own stocks of the bacterium. Alternatively, pre-existing low levels of immunity through exposure to other cross-reacting mycobacteria in the environment have been claimed to provide as much protection against TB as can BCG. Because there appears to be a correlation between the geographical location of BCG trials and the heterogeneous results, it has been suggested there may be a genetic link, and there is some experimental evidence to support this hypothesis.

**Table 11.2:** Results from clinical trials of BCG vaccine

| Country | Year | Population | | Cases of TB | | TB deaths | |
|---|---|---|---|---|---|---|---|
| | | BCG | No BCG | BCG | No BCG | BCG | No BCG |
| England | 1977 | 13 598 | 12 867 | 62 | 248 | — | — |
| India | 1980 | 88 391 | 88 391 | 505 | 499 | — | — |
| Puerto Rico | 1974 | 50 634 | 27 338 | 186 | 141 | 8 | 12 |
| USA | 1976 | 16 913 | 17 854 | 27 | 29 | — | — |

Derived from ref. [1].

Protective immunity against *M. tuberculosis* is mediated mainly by cellular rather than humoral immunity, but T cells also play a part in the pathogenesis of TB. This delicate balance presents a significant complication to vaccine development. Currently this is based on the analysis of subsets of T lymphocytes and the characterization of individual antigens or epitopes. Such an approach will probably lead to the identification of one or more protective antigen which could be combined into a cocktail of synthetic peptides, or the appropriate genes, once they have been identified or cloned and inserted into a suitable vector. Another approach is to 're-invent the wheel' by the further development of vaccines based on attenuated strains of pathogenic mycobacteria but by the application of genetic manipulation techniques. Interestingly, injection of killed BCG confers little protection in animal models.

Gastrointestinal tract infections are also associated with a very wide range of pathogens including viruses, bacteria, protozoan and worm infections. Again, antigenic diversity presents a serious problem. Many different strains of *E. coli* are associated with diarrheal disease, and there are more than 2000 members of the genus *Salmonella* distinguished by their cellular (O) and flagellar (H) antigens as used in the Kauffmann–White classification scheme. Nonbacterial gastroenteritis and diarrhea is responsible for the deaths of several million infants each year. Rotaviruses are important causes of such diseases, and these viruses infect the young of many different animals including piglets and calves. Although there appears to be a relatively small number of serotypes, human rotaviruses do not grow readily in cell culture. One approach to a rotavirus vaccine has been, in fact, to use bovine or monkey strains of the virus as heterologous vaccines to immunize human infants. Because these viruses have a double-stranded RNA genome consisting of 11 separate segments, exchange of genetic information (reassortment) occurs very readily between different rotavirus strains. This phenomenon may be used for the further development of reassorted rotavirus vaccines.

A vaccine for AIDS is clearly a paramount requirement but, more than a decade after the emergence of HIV in the human population, the problem has not been resolved. Why? Many explanations – and some excuses – may be offered, but they focus on the variety of unpleasant answers this pathogen has held in store for its human inquisitors. Much attention has been directed towards the surface glycoproteins of the virion, particularly gp120 or its progenitor, gp160. Although part of this molecule, the V3 loop, contains epitopes which elicit neutralizing antibodies, this region is hypervariable and neutralization-escape mutants readily appear. Rather more alarmingly, whether or not a neutralizing antibody is important in resistance to HIV infection *in vivo* has still not been established. In fact, some antibodies may enhance infection by one strain which differs by only one amino acid residue in the V3 loop from another strain that is neutralized by the same antibody. In relation to CMI, an alternative approach to circumvent these problems may be to elicit immune responses to internal, group-specific antigens such as the p24 core antigen. Such a rationale is based on the induction of CTL by the virion's internal nucleoprotein antigen; cellular immune responses to its NP antigen have been shown to aid recovery from influenza virus infection. However, there is evidence from cytokine profiles that there is a shift towards the predominance of a Th2-like helper cell population as HIV infection progresses. This may be related to the decline of the ability of CTL to inhibit HIV replication as disease progresses. This could be exacerbated by the progressive depletion of dendritic cells and the concomitant loss of their antigen-presentation function together with the decline in $CD4^+$ Th cells. Another relatively undeveloped aspect of vaccine design that is particularly applicable to HIV infection is the need to elicit mucosal immunity. This approach is facilitated by microencapsulation technology which is being used to facilitate both protection of the antigen and its delivery at the required site. For example, antigens encapsulated in PLDLG microspheres are protected against destruction in the gastrointestinal lumen and PLDLGs of 5–10 µm selectively stimulate mucosal immune responses. However, mucosal immunity at the epithelial surfaces of genital and rectal areas are required to protect against HIV infection. It is also relevant that certain phenotypes of HIV, the nonsyncytium-inducing, macrophage-tropic strains, may be selectively transmitted by sexual contact even though they represent a minor component of the host's total infectious load.

Although an attenuated vaccine to protect against simian AIDS has been developed, progress towards attenuated or inactivated HIV vaccines is severely limited by safety issues. It will be necessary, however, to identify appropriate epitopes which can be modified to enhance their antigenicity and ability to be cross-reactive with different virus strains. Thus, future developments are likely to move towards vectored or subunit vaccines or even DNA immunization. As indicated earlier (see Chapter 7) many putative HIV vaccines have been produced as recombinant proteins or recombinant poxviruses but their assessment in phase I trials has been based on seroconversion determined by circulating antibody titers. The further development of HIV vaccines is clearly essential, particularly to induce mucosal immunity.

Another disease of global significance is malaria, which infects one-third of the world's population, resulting in an estimated 1.5 million deaths each year. Due to increased tourism and air travel, malaria is now seen in the developed world where its unexpected presence may delay diagnosis with fatal results. The climatic changes that are predicted to result from global warming could also bring to many developed countries an increased risk of malaria and other insect-transmitted infectious and parasitic diseases.

Deaths from malaria are mainly in young children (up to 5 years of age) and in nonimmune adults (possibly tourists). Young adults in endemic areas do develop clinically effective immunity although this is never complete. Also, persistent antigenic stimulation appears to be necessary because immunity wanes if a period longer than 1 year is spent away from a malaria endemic area. The reasons for such slow induction of protective immune responses and their relatively transient nature once they do develop are unknown. They may be due to the wide diversity of antigenic stimuli offered by the parasite, only some of which will actually evoke the protective immune responses that are required; the need for different humoral responses to attack each extracellular stage (sporozoites, merozoites, gametocytes and gametes); antigenic variation during the blood stage; and the need for cell-mediated immune responses to tackle the intracellular stages.

Animal models have shown malaria vaccines are attainable but human vaccines are still in field trials. Subunit vaccines are most favored because it is important with this parasite to try to identify protective epitopes for B-cell and T-cell responses and to eliminate other diversionary antigenic stimuli. Also, over-induction of cytokines during natural infection appears to cause damage in the brain, lungs, liver and other organs in severe malaria. Early trials using a synthetic peptide or recombinant peptide based on the major surface protein of the sporozoite have not been very successful. A more promising approach has been based on a chimeric molecule SPf(66)n which is a chemically synthesized 45 amino acid peptide derived from four different proteins of the sporozoite. Phase III efficacy trials are being carried out with children aged between 1 and 5 years in the Kilombero valley in Tanzania (Africa).

Vaccines for other parasitic diseases also present formidable problems, but for different reasons: the structural complexity of both protozoan and worm parasites; their complex life cycles; various mechanisms of immune evasion they have adopted; and how to demonstrate the efficacy of a putative parasite vaccine when large numbers of people will need to be vaccinated and followed for many years.

Recent years have seen the rapid development of techniques for the production in large quantities of monoclonal antibodies with human characteristics that can be made with virtually any chosen specificity. Their passive administration may be used both therapeutically and prophylactically to control a wide variety of infectious and parasitic diseases, particularly chronic, persistent or recurrent infections.

## 11.6 Sterile immunity: is it possible?

To date, the vaccine strategy in many communicable diseases has not been to prevent infection but to achieve, through secondary immune responses, the eventual total elimination of the pathogen before disease is caused. However, such so-called sterile immunity is difficult to achieve in certain cases. This applies particularly to latent virus infections established by herpesviruses, but the most insidious example is HIV, which is able to integrate its genome into host cell chromosomes. Such covert presence clearly remains undetected, and it is difficult to see how a vaccine by artificial immunization can achieve what is impossible by natural immunity. In such cases, the alternative approach must be to vaccinate against the disease by preventing virus dissemination after reactivation although such immunity must be maintained for a lifetime. This will require a therapeutic rather than prophylactic use of vaccines and such measures have already been assessed in clinical trials. Vaccination with recombinant glycoprotein D of HSV type 2 has been used to treat recurrent genital herpes. A phase I evaluation of the safety and immunogenicity of vaccination with recombinant gp160 has also been carried out in patients with early HIV infection. In fact, the control of HIV infection may be better achieved by a combination of chemotherapy and immunotherapy; reduction of virus load by drug treatment accompanied by artificial stimulation of immune responses by vaccination. Encouragement to adopt such approaches may be taken from Louis Pasteur's therapeutic use of his rabies vaccine and the more recent demonstration that onchocerciasis patients treated repeatedly with ivermectin showed reversal of parasite-mediated immunosuppression and activation of parasite-specific Th1-type responses.

## 11.7 New and re-emerging infectious diseases

The development of vaccines must be a continuous process because new infectious diseases continue to appear. The most notorious example is AIDS, caused by a virus recently introduced into the human population probably from another primate host. The Hantaan virus, which causes pulmonary disease, is another recent discovery (see Chapter 6). In addition, old diseases, such as cholera, appear in new places such as South and Central America. STDs – gonorrhea, syphilis and chlamydial infections – are becoming increasingly more common in developed countries. In developing countries human settlement in new geographical areas and the introduction of novel agricultural techniques will change natural habitats in ways that may release new human pathogens. Other changes in human patterns, such as the increasing number of children who are in day care centres, have seen increased incidence of diarrhea caused by *G. lamblia, Shigella* spp. and rotaviruses. Modern farming methods in developed countries have also been responsible for greater numbers of diseases due to food-borne infections with *Listeria monocytogenes, E. coli* and *Salmonella enteritidis.*

The design of new vaccines is clearly essential but its achievement will be a real test of modern immunological and molecular biological skills.

## References

1. Colditz, G.A., Brewer, T.F., Berkey, C.S., Wilson, M.E., Burdick, E., Fineberg, H.V. and Mosteller, F. (1994) *J. Am. Med. Assoc.,* **271,** 698.

# Appendix A. Glossary

**Acquired (adaptive) immunity:** the part of the immune system that responds in a specific manner to foreign antigen.

**Adjuvant:** an agent administered with an antigen which induces heightened immune responses.

**Allele:** short for allelomorph, meaning alternative form of the same gene.

**Antibody:** immunoglobulin synthesized by B lymphocytes in response to antigen and able to bind specifically with that antigen.

**Antigen:** a molecule, usually a protein, that causes a specific immune response.

**Antigen presentation:** a number of different leukocytes (antigen-presenting cells) express antigen on their cell surface which can stimulate either B or T cells.

**Autocrine:** a mediator that acts on the cell by which it was produced.

**Autonomously replicating sequences (ARSs):** segments of DNA which, when inserted into a circular plasmid, confer the ability to replicate extra-chromosomally in yeast.

**B lymphocyte:** a cell that makes immunoglobulin.

**cDNA clone:** a recombinant molecule containing the double-stranded DNA copy of an mRNA sequence.

**cDNA library:** a collection of cDNA clones, normally maintained as plasmids in bacterial cells..

**Cell-mediated immunity:** immune responses directed against intracellular pathogens in which antibody plays little part. Mediated by cytotoxic T cells, natural killer (NK) cells and lymphokine-activated killer (LAK) cells.

**Chromosome:** structure in the nucleus on which genes are located. Humans have 22 paired autosomal chromosomes, and one X and one Y chromosome (male) or two X chromosomes (female).

**Cluster designation (CD):** a number given by an international workshop to a membrane structure recognized by two or more monoclonal antibodies. Usually defines a glycoprotein membrane receptor.

**Codon:** a triplet of nucleotides in an mRNA molecule that specifies the insertion of a particular amino acid into the growing polypeptide chain.

**Complement pathway:** an enzyme cascade system which is activated when certain classes of antibody recognize and bind surface antigen. Activated complement can lyse cell membranes or virus envelopes.

**COS (COS7) cells:** CV-1 monkey kidney cells transformed with an origin-defective form of SV40. The cells express SV40 T antigen and will replicate plasmids containing the SV40 origin of replication when introduced into the cells. As the SV40 used to construct the cells is origin defective, T antigen will not bind and replicate the endogenous SV40 genome.

**Cosmid:** a plasmid vector that contains *cos* sites and which can therefore be packaged into pseudo-viral particles.

**Cross-hybridization:** the annealing of two nucleic acid molecules which are not perfectly complementary.

**Cytokines:** group name for a complex and heterogenous mixture of peptide mediators released by a number of cell types that modulate cell function in a variety of ways. They often have a role in controlling cell growth and differentiation.

**Deletion series:** a set of clones, all derived from the same initial recombinant, but in which the insert lacks sequences at one of its ends because of treatment with an exonuclease.

**Denaturation:** a process whereby the two strands of a double-stranded nucleic acid molecule come apart as a result of heating or exposure to alkaline conditions.

**Discontinuous/conformational epitope:** epitope determined by the three-dimensional folding of an antigen molecule.

**DNA polymerase:** an enzyme which synthesizes double-stranded DNA from single-stranded DNA.

**DNA sequence homology:** the degree of identity between DNA sequences.

**ELISA:** enzyme-linked immunosorbent assay. Methods using enzyme-conjugated species-specific antibodies as probes to detect the binding of test antibodies to antigen. The binding of antibody is quantified by the degree of enzyme-dependent color development of a substrate.

**Endonuclease:** a nuclease which cuts a nucleic acid molecule by cleaving between two internal residues.

**Enhancers:** eukaryotic gene regulatory elements that can confer tissue specificity of expression and which can activate gene transcription even when situated great distances away from the cap site of the gene.

**Epitope:** region of a protein recognized by the antigen recognition structures of either immunoglobulin or the T-cell antigen receptor. *See also* Discontinuous/conformational epitope, Epitope mapping *and* Linear epitope.

**Epitope mapping:** method by which epitopes recognized by T or B cells can be determined.

**Exon:** a section of a gene or of an mRNA precursor which is incorporated into the mature mRNA.

**Expression screening:** a method of screening for a specific cDNA clone where the cDNA is inserted next to a promoter active in the host cell and an immunoassay or bioassay is used to detect the required clone.

**Expression vector:** a cloning vector which will enable a foreign gene to be expressed in the host organism.

**Gene:** sequence of nucleotides in the DNA which encodes a single polypeptide. Genes contain both exons (coding sequences) and introns (noncoding sequences).

**Gene deletion:** natural or experimental removal of all or part of a gene sequence.

**Gene exons:** nucleotide sequences of a gene which encode for the expressed product.

**Gene fusion:** a DNA segment containing parts of different genes (e.g. the promoter of one gene and the coding region of another gene).

**Gene introns:** nucleotide sequences of a gene which do not encode for expressed product. Introns are removed during mRNA splicing.

**Gene mutation:** event or process whereby the nucleotide sequence of a gene is changed. This may occur 'naturally' in a random manner as a result of irradiation, chemical or other environmental effects. Alternatively, it can be engineered specifically by methods such as site-directed mutagenesis.

**Gene transfection:** methods in which a new or 'foreign' gene is introduced into the cell.

**Genome:** the sum of the genetic information necessary to specify the formation of a living organism or a virus.

**Genomic clone:** a recombinant molecule containing genomic DNA. Normally the term refers to a clone containing a gene and a variable amount of flanking DNA depending upon the method of cloning used.

**Genotype:** (a) the genetic constitution of an individual; (b) the types of alleles found at a locus in an individual.

**Glycosylation:** the addition of carbohydrate to proteins.

**Granulocytes:** neutrophil, basophil or eosinophil white blood cells.

**Granuloma (plural granulomata):** a small mass of granulation tissue.

**Haplotype:** description of the types of alleles found at linked loci on a single chromosome.

**Hemopoiesis:** the generation of blood cells from bone marrow stem cells.

**Heterozygous:** the possession by an individual of different alleles at the same locus on each of a pair of chromosomes.

**Human leukocyte antigens (HLA):** receptors (class 1 and class 2) expressed on the surface of a variety of cells which bind antigenic peptide fragments in order to present them to the T-cell antigen receptor, so resulting in a specific (i.e. directed only against that antigen) immune response.

**Humoral immunity:** in higher vertebrates refers to immune processes that are mediated by antibody.

**Hydrophobicity profile:** profile reflecting the distribution of hydrophilic or hydrophobic amino acid residues in a protein. This profile can be used to predict which residues are likely to be exposed on the surface of a molecule and contribute to its antigenicity.

**Hypervariable regions:** regions within proteins such as immunoglobulins, histocompatibility antigens and T-cell receptors where the majority of sequence variation exists.

**Idiotype:** the antigen combining sites of antibody molecules which have unique specificity for an antigen. These are encoded by variable light and heavy immunoglobulin genes.

**Immune complex:** the combination of antigen and antibody in varying proportions. Depending on the size of complex formed, it can be soluble or precipitated.

**Immune surveillance:** a mechanism whereby the immune response constantly maintains the integrity of the body by detecting and eliminating malignant or foreign cells which do not measure up to the strict definition of what is self.

**Immunoglobulin class switching:** a process whereby immunoglobulins change from having one type of constant heavy chain to another.

**Inflammation:** a local, natural response to tissue injury or destruction. Illustrated by the wheal and flare reaction and characterized by redness, swelling, heat and pain.

**Innate immunity:** the part of the immune system which is not dependent on recognition of specific antigens for its function. Includes the complement system, phagocytosis and anti-bacterial components of saliva.

**Insert:** genomic cDNA contained within a vector.

***In situ* hybridization:** binding of a labeled cloned gene or oligonucleotide probe to a large DNA molecule, usually a chromosome. This technique can be used for gene mapping.

**Interferons:** glycoproteins that have nonspecific antiviral activity against virus-infected cells. Interferons also modulate the growth and differentiation of a variety of leukocytes. Their production is stimulated by intracellular pathogens (DNA and RNA viruses, chlamydia and protozoa), toxins and cytokines.

**Integrins:** a large family of cell surface adhesion receptors characterized by a two-chain heterodimeric molecular structure.

**Interleukins:** cytokines produced by a variety of cell types which regulate many immune functions as well as other processes.

**Intron:** part of a nuclear RNA precursor which is removed and degraded within the nucleus during mRNA processing.

**Leukotrienes:** pharmacologically active products of arachidonic acid metabolism.

**Ligand:** the structure which binds to a receptor.

**Linear epitope:** epitope determined by a continuous linear sequence of amino acid residues.

**Lymphocytes:** white blood cells of the lymphoid lineage which originate from bone marrow stem cells. B lymphocytes develop in the bone marrow and differentiate into plasma cells to produce antibodies. T lymphocytes develop in the thymus to become either T-helper cells or cytotoxic T cells.

**Lymphokine:** cytokine produced by cells of the immune system.

**Lymphoma:** a solid tumour resulting from malignant transformation in lymphoid tissue.

**Macrophage:** phagocytic cells derived from monocytes which are capable of ingesting foreign cells and particles and, after protein degradation, are able to present antigen to T lymphocytes.

**Major histocompatibility complex (MHC):** a family of highly polymorphic genes encoding a variety of cell surface glycoprotein transplantation antigens which are responsible for the presentation of peptide antigens to

T lymphocytes. They provide the mechanism by which self can be distinguished from nonself.

**Mast cells:** cells containing a variety of inflammatory substances (such as vasoactive amines, histamine and heparin) which can be released through the action of IgE antibodies.

**MCS (multi-cloning site or polylinker):** a short DNA sequence, found in most vectors in common use, which contains many closely spaced restriction enzyme cleavage sites.

**Memory T cells:** T lymphocytes that have been stimulated by antigen through their antigen receptor.

**Messenger RNA (mRNA):** a strand of nucleic acid transcribed from a gene which passes into the cytoplasm and is translated into protein on ribosomes.

**Monoclonal antibody:** antibody produced by a hybrid cell which has been made by fusing a stimulated B lymphocyte with a myeloma cell line. The fused 'hybridoma' cell is manipulated to be clonal and thus antibody with a single isotype and a single specificity is produced. The hybridoma is immortal and can be kept in continuous culture to produce limitless supplies of identical antibody.

**Monocytes:** *see* Macrophage.

**Mutations:** alterations in the nucleotide sequence of a DNA molecule.

**Naive T cells:** lymphocytes that have not encountered antigen.

**Natural killer (NK) cells:** lymphocytes, which are not B or T cells, in nonimmunized individuals that have MHC-independent cytolytic activity against tumour cells cells and virus-infected cells.

**Neutralizing antibody:** an antibody which is able to prevent virus replication.

**NYVAC:** vaccinia virus strain Copenhagen deleted in six different genomic regions to give a highly attenuated virus that cannot replicate in human cells.

**Open reading frames (ORFs):** a region of DNA sequence that does not contain one of the three stop codons and therefore has the potential to encode a protein.

**Opportunistic infections:** diseases caused in immunocompromised individuals by organisms that are usually nonpathogenic.

**Phenotype:** the appearance or other characteristics of an organism.

**Plaque:** an area of a bacterial or tissue culture monolayer where the cells are dead, or grow slowly because of infection by a virus.

**Plasma cell:** antigen-stimulated, Ig-producing B cell.

**Plasmid:** extrachromosomal, circular DNA molecules found in prokaryotes and some lower eukaryotes.

**Poly(A) tail:** a tract of A residues of approximately 100–200 nucleotides in length that is added enzymatically to the 3′ end of an mRNA.

**Polymerase chain reaction (PCR):** a method of specifically copying, and amplifying, a part of a nucleic acid chain.

**Pooled (amplified) libraries:** for long-term storage phage particles or bacterial colonies on bacteriological plates are eluted in a mixed pool, containing representatives of every different recombinant on the plate.

**Primer:** a short nucleic acid molecule which, when annealed to a complementary template strand, provides a 3′ terminus suitable for copying by a DNA polymerase.

**Probe:** a nucleic acid sequence that is complementary to part or all of a target which is to be detected by hybridization. It is usually 'tagged' by the incorporation either of radioactively labeled nucleotides or of nucleotides which are chemically modified in such a way that they can be identified immunologically.

**Prophylaxis:** preventive treatment of disease.

**Prostaglandins:** pharmacologically active products of arachidonic acid metabolism.

**Replication origin:** a segment of DNA that acts as the start site of DNA replication.

**Restriction enzymes:** enzymes which cleave double-stranded DNA into discrete pieces by cleaving at defined recognition sequences.

**Restriction map:** a physical map of a piece of DNA showing the position of cleavage of one or more restriction enzymes.

**RMAS cells:** a mutant murine lymphoma cell line containing a premature stop codon in the *TAP-2* gene, resulting in a peptide transport deficiency and destabilizing the expression of MHC class I.

**RNA polymerase:** an enzyme which copies a DNA or an RNA molecule to produce its RNA copy.

**Seroconversion:** development of serum antibodies against a specific antigen or pathogen.

**Signal peptide:** a stretch of predominantly hydrophobic amino acids which directs the polysomes to the endoplasmic reticulum during translation.

**Splicing:** the series of steps within the nucleus whereby introns are removed from the nuclear precursor and the exons are joined together to form an mRNA.

**Subcloning:** breaking a piece of cloned DNA into smaller fragments which are then cloned individually.

**Superantigens:** antigens which can activate large proportions of the T-cell repertoire through their ability to form ligands between class II molecules and V$\beta$ elements of T-cell receptors.

**Syncytia:** multinucleate giant cells produced by the fusion of a number of individual cells.

**T lymphocyte:** cells involved in cell-mediated immunity and B-lymphocyte help.

**Th2/Th1:** T-lymphocyte subsets with characteristic profiles of cytokine secretion.

**T-cell helper:** CD4-positive antigen cell which is important in both B- and T-cell responses.

**T-cell receptor:** receptor present on the surface of T cells, capable of recognizing and combining with antigens presented in the context of MHC molecules. Two types of TCR exist, one comprising $\alpha/\beta$ chains, the other $\gamma/\beta$ chains.

**T-cell repertoire:** the total available range of antigens which can be recognized by T cells in an individual.

**T-cell suppressor:** CD8-positive antigen T lymphocyte capable of modulating immune responses by suppressing other T cells.

**TCID$_{50}$:** infective dose that will produce virus replication in 50% of infected tissue culture.

**Transcription:** the process whereby a DNA molecule is copied into RNA.

**Transfer RNA (tRNA):** the intermediaries which carry amino acids to the ribosome and which direct their position of insertion into the polypeptide chain.

**Transformation (DNA transformation):** the process whereby a DNA molecule is introduced into a living cell. The term is normally reserved for those cases where the DNA molecule has the capacity to be stably maintained in the host cell.

**Transient expression:** a technique in which DNA is introduced into eukaryotic cells and its transcription is analyzed after it has reached the nucleus, but before it has integrated into the genome.

**Translation:** the process of making a peptide (protein) chain from a strand of mRNA.

**Vector:** a DNA molecule with the ability to replicate in its cognate host cell and which contains a genetic marker that allows for selection of cells that contain it.

**Western blotting:** a procedure in which antibodies are used as 'probes' to detect proteins which have been separated electrophoretically under reducing or nonreducing conditions and blotted on to membrane. Bands of immunoprecipitation can be detected by autoradiography or ELISA techniques.

**X-ray crystallography:** by manipulation of conditions, crystals of pure protein can be grown. When subjected to X-ray diffraction studies, considerable information relating to the three-dimensional structure of the molecule can be gained.

# Appendix B. Further reading

## General

Abbas, A.K., Lichtman, A.H. and Pober, J.S. (1994) *Cellular and Molecular Immunology*. W.B. Saunders, Philadelphia.

Davis, B.D., Dulbecco, R., Eisen, H.N. and Ginsberg, H.S. (1990) *Microbiology*, J.B. Lippincott, Philadelphia.

Dimmock, N.J. and Primrose, S.B. (1994) *Introduction to Modern Virology*. Blackwell Science, Oxford.

Harper, D.R. (1994) *Molecular Virology*. BIOS Scientific Publishers, Oxford.

Harlow, E. and Lane, D. (1988) *Antibodies: a Laboratory Manual*. Cold Spring Harbor Laboratory Press, Cold Spring Harbor, NY.

Hyde, J.E. (1990) *Molecular Parasitology*. Open University Press, Milton Keynes.

Jawetz, E., Melnick, J.L., Adelberg, E.A., Brooks, G.F., Butel, J.S. and Ornston, L.N. (1991) *Medical Microbiology*. Prentice–Hall, Princeton, NJ.

Kiberstis, P.A., Benditt, J.M. and Koshland, D.E. Jr (1994) (eds) Frontiers in medicine: vaccines. *Science,* **265,** 1371; 1448.

Male, D., Champion, B., Cooke, A. and Owen, M. (1991) *Advanced Immunology*. J.B. Lippincott, Philadelphia.

Mims, C.A. (1987) *The Pathogenesis of Infectious Disease*. Academic Press, London.

Mims, C.A. and White, D.O. (1987) *Viral Pathogenesis and Immunology*. Blackwell Scientific Publications, Oxford.

Mims, C.A., Playfair, J.H.L., Roitt, I.M., Wakelin, D. and Williams, R. (1993) *Medical Microbiology*. Mosby, London.

Prescott, L.M., Harley, J.P. and Klein, D.A. (1994) *Microbiology*. W.C. Brown, Oxford.

Roitt, I.M. (1994) *Essential Immunology*. Blackwell Scientific Publications, Oxford.

Roitt, I.M., Brostoff, J. and Male, D.K. (1993) *Immunology*. Mosby, London.

Sambrook, J., Fritsch, E.F. and Maniatis, T. (1989) *Molecular Cloning: a Laboratory Manual,* 2nd edn. Cold Spring Harbor Laboratory Press, Cold Spring Harbor, NY.

Schaechter, M., Medoff, G. and Eisenstein, B.L. (1993) *Mechanisms of Microbiological Disease.* Williams & Wilkins, Baltimore.

Taussig, M.J. (1984) *Processes in Pathology and Microbiology.* Blackwell Scientific Publications, Oxford.

Williams, J., Ceccarelli, A. and Spurr, N. (1993) *Genetic Engineering.* BIOS Scientific Publishers, Oxford.

## Adjuvants

Audibert, F.M. and Lise, L.D. (1993) Adjuvants – current status, clinical perspectives and future prospects. *Immunol. Today,* **14**, 281.

Khan, M.Z.I., Opdebeeck, J.P. and Tucker, I.G. (1994) Immune potentiation and delivery systems for antigens for single step immunization; recent trends and progress. *Pharmaceut. Res.,* **11**, 2.

## Chemotherapy

Burnham, G.M. (1993) Adverse reactions to ivermectin treatment for onchocerciasis. Results of a placebo-controlled double-blind trial in Malawi. *Trans. R. Soc. Trop. Med. Hyg.,* **87**, 313.

Cohen, J. (1993) Can combination therapies overcome drug resistance? *Science,* **260**, 1258.

Editorial (1994) Plague in India: time to forget the symptoms and tackle the disease. *Lancet,* **344**, 1033.

Garrod, G.L.P., Lambert, H.P. and O'Grady, F. (1992) *Antibiotic and Chemotherapy.* Churchill Livingstone, London.

Richman, D.D. (1994) Drug resistance in viruses. *Trends Microbiol.,* **2**, 401.

Soboslay, P.T., Lüder, C.G.K., Hoffman, W.H. *et al.* (1994) Ivermectin-facilitated immunity in onchocerciasis: activation of parasite-specific Th1-type responses with subclinical *Onchocerca volvulus* infection. *Clin. Exp. Immunol.,* **96**, 238.

## Cytokines

Afonso, L.C.C., Scharton, T.M., Vieira, L.Q., Wysocka, M., Trinchieri, G. and Scott, P. (1994) The adjuvant effect of interleukin-12 in a vaccine against *Leishmania major. Science,* **263**, 235.

Clemens, M.J. (1991) *Cytokines.* BIOS Scientific Publishers, Oxford.

Grabstein, K.H., Eisenman, J., Shanebek, K. *et al.* (1994) Cloning of a T cell growth factor that interacts with the β chain of the interleukin-2 receptor. *Science,* **264**, 965.

Hamblin, A.S. (1993) *Cytokines and Cytokine Receptors.* IRL Press, Oxford.

Ramsay, A.J., Husband A.J., Ramshaw, I.A., Bao, S., Matthaei, K.I., Koehler, G. and Kopf, M. (1994) The role of interleukin-6 in mucosal IgA antibody responses *in vivo. Science,* **264,** 561.

Romagnani, S. (1994) Lymphokine production in disease states. *Ann. Rev. Immunol.,* **12,** 227.

Taub, D.D. and Oppenheim, J.J. (1994) Chemokines, inflammation and the immune response. *Ther. Immunol.,* **1,** 224.

## Elimination of infectious diseases

Evans, A.S. (1983) Criteria for control of infectious diseases with poliomyelitis as an example. *Prog. Med. Virol.,* **30,** 141.

Fenner, F., Henderson, D.A., Arita, I., Ježek, Z. and Ladnyi, I.D. (1988) *Smallpox and its Eradication.* World Health Organization, Geneva.

## Herd immunity

Anderson, R.M. and May, R.M. (1990) Immunisation and herd immunity. *Lancet,* **335,** 641.

## Immune responses

Clark, E.A. and Ledbetter, J.A. (1994) How B and T cells talk to each other. *Nature,* **367,** 425.

Engelhard, V.H. (1994) Structure of peptides associated with class I and class II MHC. *Ann. Rev. Immunol.,* **12,** 181.

Schmid, S.L. and Jackson, M.R. (1994) Making class II presentable. *Nature,* **369,** 103.

## Immunotherapy

Co, M.S., Deschamps, M., Whitley, R.J. and Queen, C. (1991) Humanized antibodies for antiviral therapy. *Proc. Natl Acad. Sci. USA,* **88,** 2869.

Redfield, R.R., Birx, D.L., Ketter, N. *et al.* and the Military Medical Consortium for Applied Retroviral Research (1991) A phase I evaluation of the safety and immunogenicity of vaccination with recombinant gp160 in patients with early human immunodeficiency virus infection. *New Engl. J. Med.,* **324,** 1677.

Riddell, S.R., Watanabe, K., Goodrich, J.M., Li, C.R., Agha, E., and Greenberg, P.D. (1992) Restoration of viral immunity in immunodeficient humans by the adoptive transfer of T cell clones. *Science,* **257,** 238.

Straus, S.E., Corey, L., Burke, R.L. *et al.* (1994) Placebo-controlled trial of vaccination with recombinant glycoprotein D of herpes simplex virus type 2 for immunotherapy of genital herpes. *Lancet,* **343,** 1460.

Winter, G. and Harris, W.J. (1993) Humanized antibodies. *Immunol. Today,* **14**, 243.

## Infectious diseases of the future

Murphy, F.A. (1994) New, emerging and reemerging infectious diseases. *Adv. Virus Res.,* **43**, 1.

## Live attenuated vectors

Almond, J.W. and Burke, K.L. (1990) Poliovirus as a vector for the presentation of foreign antigens. *Sem. Virol.,* **1**, 11.

Andino, R., Silvera, D., Suggett, S.D., Achacoso, P.L., Miller, C.J., Baltimore, D. and Feinberg, M.B. (1994) Engineering poliovirus as a vaccine vector for the expression of diverse antigens. *Science,* **265**, 1448.

Binns, M.M. and Smith, G.L. (1992) (eds) *Recombinant Poxviruses.* CRC Press, Boca Raton, FL.

Dougan, G. (1994) The molecular basis for the virulence of bacterial pathogens – implications for oral vaccine development. *Microbiology UK,* **140**, 215.

Khan, C.M.A., Villarealramos, B., Pierce, R.J. *et al.* (1994) Construction, expression and immunogenicity of the *Schistosoma mansoni* p-28 glutathione-S-transferase as a genetic fusion to tetanus toxin fragment-C in a live ARO attenuated vaccine. *Proc. Natl Acad. Sci. USA,* **91**, 11261.

Langermann, S., Palaszynski, S.R., Burlein, J.E., Koenig, S., Hanson, M.S., Briles, D.E. and Stover, C.K. (1994) Protective humoral response against pneumococcal infection in mice elicited by recombinant bacille Calmette-Guérin vaccines expressing pneumococcal surface protein-A. *J. Exp. Med.,* **180**, 2277.

Langermann, S., Palaszynski, S.R., Sadziene, A., Stover, C.K. and Koenig, S. (1994) Systemic and mucosal immunity induced by BCG vector expressing outer-surface protein-A of *Borrelia burgdorferi. Nature,* **372**, 552.

Mackett, M. (1994) Vaccinia virus recombinants: expression vectors and potential vaccines. *Animal Cell Biotechnology,* Vol. 6. Academic Press, New York.

Natuk, R.J., Lubeck, M.D., Chanda, P.K. *et al.* (1993) Immunogenicity of recombinant adenovirus–human immunodeficiency virus vaccines in chimpanzees. *AIDS Res. Hum. Retrovir.,* **9**, 395.

Tacket, C.O., Losonsky, G., Lubeck, M.D., Davis, A.R., Mizutani, S., Horwith, G., Hung, P., Edelman. R. and Levine, M.M. (1992) Initial safety and immunogenicity studies of an oral recombinant adenohepatitis-B vaccine. *Vaccine,* **10**, 673.

## Pathogenesis

Bertoletti, A., Sette, A., Chisari, F.V., Penna, A., Levrero, M., de Caril, M., Flaccadori, F. and Ferrari, C. (1994) Natural variants of cytotoxic epitopes are T-cell receptor antagonists for antiviral cytotoxic T cells. *Nature,* **369**, 407.

Brown, P. (1992) How does HIV cause AIDS? *New Scientist,* **135,** 31.

Fields, B.N. (1994) AIDS: time to turn to basic science. *Nature,* **369,** 95.

Leigh Brown, A.J. (1991) Sequence variability in human immunodeficiency viruses: pattern and process in viral evolution. *AIDS,* **5** (suppl. 2), S35.

Maizels, R.M., Bundy, D.A., Selkirk, M.E., Smith, D.F. and Anderson, R.M. (1993) Immunological modulation and evasion by helminth parasites in human populations. *Nature,* **365,** 797.

Marrack, P. and Kappler, J. (1994) Subversion of the immune system by pathogens. *Cell,* **76,** 323.

McGuire, W., Hill, A.V.S., Allsopp, C.E.M., Greenwood, B.M. and Kwjatkowski, D. (1994) Variation in the TNF-α promoter region associated with susceptibility to cerebral malaria. *Nature,* **371,** 508.

McIntosh, K. and Fishaut, J.M. (1980) Immunopathological mechanisms in lower respiratory tract disease of infants due to respiratory syncytial virus. *Prog. Med. Virol.,* **26,** 94.

McMichael, A. (1993) Natural selection at work on the surface of virus-infected cells. *Science,* **260,** 1771.

Nunn, P.D., Elliot, A.M. and McAdam, K.P.J.W. (1994) Impact of human immunodeficiency virus on tuberculosis in developing countries. *Thorax,* **49,** 511.

Oehen, S., Hengartner, H. and Zinkernagel, R.M. (1991) Vaccination for disease. *Science,* **251,** 195.

## Smallpox

Fenner, F., Henderson, D.A., Arita, I., Jĕzek, Z. and Ladnyi, I.D. (1988) *Smallpox and its Eradication.* World Health Organization, Geneva.

Jenner, E. (1798) *An Inquiry into the Cause and Effects of the Variolae Vaccinae, a Disease, Discovered in some of the Western Counties of England, Particularly Gloucestershire, and Known by the Name of the Cowpox.* Sampson Low, London. Reprinted (1958) in *Classics of Medicine and Surgery* (C.N.B. Camac, ed.). Dover, New York, p. 213.

## Subunit vaccine production

Bathhurst, I.C. (1994) Protein expression in yeast as an approach to production of recombinant malaria antigens. *Am. J. Trop. Med.,* **50,** SS20.

De Wilde, M., Cabezon, T., Harford, H., Rutgers, T., Simoen, E. and van Wijnendaele, F. (1983) Production in yeast of hepatitis B surface antigen by R-DNA technology. *Dev. Biol. Standards,* **59**, 99.

King, L.A. and Possee, R.D. (1992) *The Baculovirus Expression System: a Laboratory Guide.* Chapman and Hall, London.

O'Reilly, D.R., Miller, L.K. and Luckow, V.A. (1992) *Baculovirus Expression Vectors. A Laboratory Manual.* W.H. Freeman, New York.

Shouval, D., Ilan, Y., Adler, R., Deepen, R., Panet, A., Evenchen, Z., Gorecki, M. and Gerlich, W.H. (1994) Improved immunogenicity in mice of a mammalian cell-derived recombinant hepatitis B vaccine containing preS1 and preS2 antigens as compared with conventional yeast derived vaccines. *Vaccine,* **12**, 1453.

Straus, S.E., Corey, L., Burke, R.L. *et al.* (1994) Placebo-controlled trial of vaccination with recombinant glycoprotein D of herpes simplex virus type 2 for immunotherapy of genital herpes. *Lancet,* **343**, 1460.

## T-cell subsets

Cox, F.E.G. and Liew, F.Y. (1992) T-cell subsets and cytokines in parasitic infections. *Immunol. Today,* **13**, 445.

Grazio, C., Pantelo, G., Gantt, K.R., Fortin, J-P., Demarest, J.F., Cohen, O.J., Sékaly, R.P. and Fauci, A.S. (1994) Lack of evidence for the dichotomy of $T_H1$ and $T_H2$ predominance in HIV-infected individuals. *Science,* **265**, 248.

Maggi, E., Mazzetti, M., Ravina, A. *et al.* (1993) Different roles of $\alpha\beta$ and $\gamma\delta$ T cells in immunity against an intracellular bacterial pathogen. *Nature,* **365**, 53.

Prete, G. and Romagnani, S. (1994) Ability of HIV to promote a $T_H1$ to $T_H0$ shift and to replicate preferentially in $T_H2$ and $T_H0$ cells. *Science,* **265**, 244.

## Vaccination practice in developed countries

Department of Health, Welsh Office, Scottish Office Home and Health Department, DHSS (Northern Ireland) (1992) *Immunisation against Infectious Diseases.* HMSO, London.

Fedson, D.S. (1994) Adult immunization. Summary of the National Vaccine Advisory Board. *J. Am. Med. Assoc.,* **272**, 778.

Hinman, A.R. and Orenstein, W.A. (1990) Immunisation practice in developed countries. *Lancet,* **335**, 707.

## Vaccination practice in developing countries

Hall, A.J., Greenwood, B.M. and Whittle, H. (1990) Practice in developing countries. *Lancet,* **335**, 774.

## Vaccines

### *Cholera*

Mekalanos, J.J. and Sadoff, J.C. (1994) Cholera vaccines: fighting an ancient scourge. *Science,* **265,** 1387.

### *Cytomegalovirus*

Plotkin, S.A. (1994) Vaccines for varicella-zoster virus and cytomegalovirus: recent progress. *Science,* **265,** 1383.

### *H. influenzae (Hib)*

Booy, R., Hodgson, S., Carpenter, L. *et al.* (1994) Efficacy of *Haemophilus influenzae* type b conjugate vaccine PRP-T. *Lancet,* **334,** 362.

### *Hepatitis B virus*

Ellis, R.W. (1993) (ed.) *Hepatitis B Virus Vaccines in Clinical Practice.* Marcel Dekker, New York.

### *HIV (AIDS)*

Blower, S.M. and McLean, A.R. (1994) Prophylactic vaccines, risk behavior change, and the probability of eradicating HIV in San Francisco. *Science,* **265,** 1451.

Cease, K.B. and Berzofsky, J.A. (1994) Towards a new vaccine for AIDS: the emergence of immunobiology-based vaccine development. *Ann. Rev. Immunol.,* **12,** 923.

Cohen J. (1993) (ed.) AIDS: the unanswered questions. *Science,* **260,** 1253.

Koup, R.A. and Ho, D.D. (1994) Shutting down HIV. *Nature,* **370,** 416.

Moore, J. and Anderson, R. (1994) The WHO and why of HIV vaccine trials. *Nature,* **372,** 313.

Nye, K.E. and Parkin, J.M. (1994) *HIV and AIDS.* BIOS Scientific Publishers, Oxford.

Pantelo, G., Demarest, J.F., Soudeyns, H., Adelsberger, J.W., Borrow, P., Saag, M.S., Shaw, G.M., Sekay, R.F. and Fauci, A.S. (1994) Major expansion of CD8[+] T cells with a predominant Vβ usage during the primary immune response to HIV. *Nature,* **370,** 463.

### *Influenza*

Nichol, K.L., Margolis, K.L., Wourenma, J. and Von Sternberg, T. (1994) The efficacy and cost effectiveness of vaccination against influenza among elderly persons living in the community. *New Engl. J. Med.,* **331,** 778.

## Malaria

Alonso, P.L., Armstrong Scellenberg, J.R.M., Masanja, H. *et al.* (1994) Randomised trial of efficacy of SPf66 vaccine against *Plasmodium falciparum* malaria in children in southern Tanzania. *Lancet,* **344,** 1175.

Noya, G.O., Berti, Y.G., de Noya, B.A. *et al.* (1994) A population-based clinical trial with the SPf66 synthetic *Plasmodium falciparum* malaria vaccine in Venezuela. *J. Infect. Dis.,* **170,** 396.

Nussenzweig, R.S. and Long, C.A. (1994) Malaria vaccines: multiple targets. *Science,* **265,** 1381.

## Measles

Katz, S.L. and Gellin, B.G. (1994) Measles vaccine: do we need new vaccines or new programs? *Science,* **265,** 1391.

## Pertussis

Miller, E., Ashworth, L.A.E., Robinson, A., Waight, P.A. and Irons, L.I. (1991) Phase II trial of whole-cell pertussis vaccine vs. an acellular vaccine containing agglutinogens. *Lancet,* **337,** 70.

## Pneumococcus

Siber, G.R. (1994) Pneumococcal disease; prospects for a new generation of vaccines. *Science,* **265,** 1385.

## Poliomyelitis

Beale, A.J. (1990) Poliovaccines: time for a change in immunisation policy? *Lancet,* **335,** 839.

Patriarca, P.A., Foege, W.H. and Swartz, T.A. (1994) Progress in polio eradication. *Lancet,* **342,** 1461.

## Respiratory syncytial virus

Doherty, P. (1994) Vaccines and cytokine-mediated pathology in RSV infection. *Trends Microbiol.,* **2,** 148.

Hall, C.B. (1994) Prospects for a respiratory syncytial virus vaccine. *Science,* **265,** 1393.

Welliver, R.C., Tristram, D.A., Batt, K., Sun, M., Hogerman, D. and Hildreth, S. (1994) Respiratory syncytial virus-specific cell mediated immune responses after vaccination with a purified fusion protein subunit vaccine. *J. Infect. Dis.,* **170,** 425.

### Rotavirus

Glass, R.I., Gentsch, J. and Smith, J.C. (1994) Rotavirus vaccines; success by reassortment? *Science,* **265,** 1389.

### Tuberculosis

Colditz, G.A., Brewer, T.F., Berkey, C.S., Wilson, M.E., Burdick, E., Fineberg, H.V. and Mosteller, F. (1994) Efficacy of BCG vaccine in the prevention of tuberculosis. *J. Am. Med. Assoc.,* **271,** 698.

Kaufmann, S.H.E. and Young, D.B. (1992) Vaccination against tuberculosis and leprosy. *Immunobiology,* **184,** 208,

Silva, C.L., Silva, M.F., Pietro, R.C.L.R. and Lowrie, D.B. (1994) Protection against tuberculosis by passive transfer of T-cell clones recognizing mycobacterial heat-shock protein 65. *Immunology,* **83,** 341.

### Varicella-zoster virus

Plotkin, S.A. (1994) Vaccines for varicella-zoster virus and cytomegalovirus: recent progress. *Science,* **265,** 1383.

### DNA

Ulmer, J.B., Donnelly, J.J., Parker, S.E. *et al.* (1993) Heterologous protection against influenza by injection of DNA encoding a viral protein. *Science,* **259,** 1745.

# Index

05